PRÉSENCE DE DEMAIN
OUR DEBT TO THE FUTURE

EDITED BY

E. G. D. MURRAY

F.R.S.C.

AT ITS ANNUAL MEETING in 1957, the Royal Society of Canada, celebrating the seventy-fifth anniversary of its foundation, departed from the accustomed pattern of its meetings. Instead of assembling in separate sections, Fellows from each Section of the Society were asked to contribute to a conspectus, focused by their specialized knowledge and trained discrimination, to reveal to the Society and to others certain trends and tendencies in Canada. Subjects and contributors are: "These Seventy-Five Years" (Presidential Address by W. A. Mackintosh); "The Roles of the Scientist and the Scholar in Canada's Future" (W. A. Mackintosh, David L. Thomson); "The Penalties of Ignorance of Man's Biological Dependence" (E. G. D. Murr , K. W. Neatby, I. McT. Cowan, G. H. Ettinger, R. H. Manske); "The Social Impact of Modern Technology" (N. A. M. MacKenzie, V. W. Bladen, E. W. R. Steacie, W. H. Watson); "Our Economic Potential in the Light of Science" (H. C. Gunning, J.E. Hawley, L. M. Pidgeon, B. S. Keirstead, Maurice Lamontagne); "Human Values and the Evolution of Society" (G.-H. Lévesque, T. W. M. Cameron, A. S. P. Woodhouse, R. Elie, Roy Daniells); "Let Us Look to Our Human Resources" (F. H. Underhill, J. K. W. Ferguson, L.-P. Dugal, W. B. Lewis).

The volume is further prefaced by the address given by His Excellency the Right Honourable Vincent Massey, Governor-Generl of Canada, "The Weighing of Ayre".

ROYAL SOCIETY OF CANADA
"STUDIA VARIA" SERIES

1. *Studia Varia: Literary and Scientific Papers—Etudes littéraires et scientifiques* (1956). Edited by E. G. D. MURRAY
2. *Our Debt to the Future: Symposium Presented on the Seventy-fifth Anniversary, 1957—Présence de demain: Colloque présenté au Soixante-quinzième Anniversaire, 1957.* Edited by E. G. D. MURRAY

Our Debt to the Future

ROYAL SOCIETY OF CANADA

Symposium presented on the
Seventy-fifth Anniversary
1957

Présence de demain

SOCIÉTÉ ROYALE DU CANADA

Colloque présenté au
Soixante-quinzieme Anniversaire
1957

EDITED BY

E. G. D. MURRAY, F.R.S.C.

PUBLISHED FOR THE SOCIETY

BY UNIVERSITY OF TORONTO PRESS

1958

PREFACE

THE 1957 MEETING marked the seventy-fifth year of the Royal Society of Canada and the Council most fittingly designed that it should be in Ottawa, where the first meeting of the Society was held.

To distinguish the occasion it was decided to depart from the accustomed pattern of meetings by having Fellows from each of the Sections of the Society contribute to a conspectus, focused by their specialized knowledge and trained discrimination, to reveal to the Society and others particular appearances of trends and tendencies in Canada.

In this century there are giants in some branches of learning and they, "mounted on the shoulders of giants," thus extend our view that we may hope to see better where we have come from and whither we may go. But so particular in language and method is some of this learning that interpreters are necessary for the rest of us to gain the understanding required for full use of available knowledge. Then, too, enthusiasm must be judiciously governed, as well for those who peer into the future as for others who "lean too much upon the past and make too much of ancient wisdom." And boundaries must be crossed, to integrate information from every source, to disentangle motives, to recognize restrictions and to view perspectives, that a composite can be made to give inspiration to purpose. For there is an urgent restlessness thrusting man the history maker ever onward, fearful lest he hear once again the mocking echo of the scholar's song:

> Temps s'en va
> Et rien n'ai fait
> Temps s'en vient
> Et ne fais rien.

The purpose of the Symposium warranted it taking up the greater part of the programme of the 1957 meeting of the Society and its ambition is expressed in the title "Our Debt to the Future." But its achievement depended not merely on the expression of all shades of opinion on a diversity of problems, some as old as history and others new, but also on the selection and grouping of speakers. With the adoption of the plan for the meeting the difficulties emerged and Professor W. H. Watson was appointed to cope with them. His success is attested by the resolution thanking him at the General Meeting of the Society.

Professor Watson, whose clear conception of the far-reaching possibilities of this discussion of Canadian affairs and whose inspiring persuasiveness impelled participation, brought together the groups of speakers to formulate the scope and nature of the undertaking by each of the panels. The general arrangement of the Symposium was effected by a Programme Committee to assist the President of the Royal Society, W. A. Mackintosh, and the Vice-President, T. W. M. Cameron: Guy Sylvestre (I), F. H. Underhill (II), H. S. Bostock (IV), W. H. Cook (V), and W. H. Watson (III), Chairman, in co-operation with the Programme Committees of the Sections. With thoughtful preparation an integrated intellectual venture was realized. Limited time prevented exhaustive surveys in any subject and highly technical arguments were avoided in order to present principles and specialist opinion as simply as possible, to interest everyone concerned with the contribution scientists and scholars can make to our future development.

The use in all countries of recent discoveries in physical sciences to influence diplomacy, legislation and public opinion causes concern in informed circles, lest those exercising these powers are not always sufficiently instructed and guided. Furthermore, recognition of the need for wise diffusion in understandable terms of expert interpretation of matters of moment should be extended, it seems to some of us, to the various branches of the biological sciences, the social studies and the humanities. Perhaps the experiment undertaken here will bring encouragement leading to a wider exploration and exposition. Perhaps there will arise an emergency to correct the political threat to submerge learning and scholarship in a technological servitude.

This anniversary of the Royal Society of Canada provided an occasion of merit to allow His Excellency the Governor-General, The Right Honourable Vincent Massey, to address the Society in words of high significance and for him to permit his address to be published as a keynote of this volume. To this is joined the address by Dr. W. A. Mackintosh, President of the Royal Society of Canada, which in itself "gaining some deeper insight into present and future" points up the purpose and significance of this Symposium by clearly defining the new Canadian era and the challenge provided by the opportunities of the future.

Except for these two substantial additions, the Symposium is published as it was designed and given. Acknowledgement and special thanks are due to the Editorial Department of the University of Toronto Press, for without its taking over unexpected responsibility it is unlikely that this volume could have appeared by the date set by the Society.

E. G. D. MURRAY

CONTENTS

PRESENCE DE DEMAIN
OUR DEBT TO THE FUTURE

"THE WEIGHING OF AYRE"

HIS EXCELLENCY THE GOVERNOR-GENERAL, THE RIGHT
HONOURABLE VINCENT MASSEY, C.H., F.R.S.C.

I MUCH APPRECIATE the more than kind welcome which you have given me. I am very happy indeed to be with you on this occasion.

Before I proceed with my remarks, I wish to make an announcement, which I do at the President's request. I have been asked to tell you that His Royal Highness, Prince Philip, Duke of Edinburgh, has been pleased to accept the invitation of the Royal Society of Canada to become an Honorary Fellow. I know this information will be received with the greatest pleasure and satisfaction by all those who belong to this distinguished body and by the Canadian people.

As I said, I am delighted to foregather with you this evening. I have been making so many speeches of late, I am reminded sometimes of a passage in Shakespeare's *Much Ado about Nothing.* You will recall what the candid Beatrice says to Benedick, "I wonder that you will still be talking, Signior Benedick: nobody marks you." In moments of depression, I am haunted with the idea that the audience marks only my indiscretions.

To be serious, like everyone here, I am sensible of the dignity as well as of the enjoyment of this occasion. I am here this evening not only as an occupant of my present post, but also I am happy that I have the privilege of coming in another capacity—as one of your Honorary Fellows. Whatever the role may be in which I appear this evening, I deeply appreciate the honour you have done me in inviting me to address you when, on your Seventy-Fifth Anniversary, you gather in our capital city. You have given me a heavy responsibility, but I could not well refuse it.

No one need be reminded that it was one of my distinguished predecessors who founded this Society three-quarters of a century ago in the hope that it would help the young nation to grow in unity and understanding, as well as in intellectual power and distinction. Lord Lorne, as he then was, did not find that his initiative met with universal approval —acts of imagination seldom do. A powerful newspaper of the day refused "to hold the Governor-General responsible for a project so absurd," and observed, in elegant language, that the new body would be "a mutual admiration society of nincompoops"! (I have much sympathy for those who, with commendable motives, initiate such crack-pot schemes.) It is interesting to read in Professor McNutt's book on Lord Lorne's régime, that his determination to proceed with his idea was strengthened by his discovery of the fact than an American institution had been permitted to collect Indian relics on Canadian soil. Since the Society came into being in 1882, its Fellows have done much to direct the attention of the Canadian people to our own national life—its precious past, its stimulating present and its boundless future. Although your outlook, like that of any body of true scholars, has never been narrowly nationalistic, you have helped to make us proud of being Canadian and proud, too, of the contribution which you made on behalf of Canada to the world as a whole.

Mais à cette occasion je m'en voudrais de ne pas souligner le fait que Lord Lorne en fondant son noble projet a beaucoup tiré d'une culture d'origine française qui, pendant plus de cent cinquante ans, fut la seule au Canada. Cette première tradition canadienne n'a pas toujours été reconnue et appréciée à sa juste valeur.

Mais cela est chose du passé. D'un passé qui a toujours su trouver, je le dis avec fierté, des canadiens de langue anglaise anxieux de reconnaître avec Pierre Chauveau, premier vice-président de la Société, qu' «Il y a longtemps, bien longtemps, que l'on fait de nobles efforts pour la culture de l'esprit humain, sur les rives du Saint-Laurent.»

Nous devons beaucoup aux premiers citoyens qui dès leur arrivée en notre pays se sont donnés aux choses de l'esprit. Notre pays, alors tout jeune, a connu, grâce à eux, un départ intellectuel dont il est resté marqué. Certes nos fondateurs n'ont pas ignoré les richesses matérielles qu'un Canada du dix-septième siècle avait en abondance, mais en hommes sages ils se sont préoccupés avant tout de sauvegarder jalousement ce trésor infiniment plus précieux qu'ils apportaient avec eux : la tradition culturelle et spirituelle de la France.

Cette culture le Canada français l'a conservée à travers les générations avec courage et augmentée avec vigueur. Nous avons aujourd'hui autour de nous des héritiers de cette culture et ils lui font honneur. Les lettres françaises jouissent maintenant parmi nous de l'admiration et du respect qu'elles se sont justement attirés et c'est avec plaisir que je me fais l'interprète des canadiens d'expression anglaise en vous disant notre reconnaissance pour l'enrichissement qu'elles apportent à notre intellect et l'espoir que nous entretenons de les voir s'épanouir de plus en plus pour le bénéfice de la nation toute entière.

Et c'est mon privilège ce soir d'offrir à vous tous de la Société Royale mes félicitations les plus sincères pour l'excellent travail que vous avez accompli au cours des années et de vous souhaiter succès et bonheur dans l'accomplissement des devoirs qui sont les vôtres.

May I say again that it is my privilege tonight to give you of the Royal Society, as a body, my warmest congratulations on the great work that you have done over the years, and to wish you good fortune and happiness in your discharge of the immense responsibilities which face you today.

These things I say in the official capacity in which I am your guest tonight. But I would like to think that your invitation to me is also personal and that, like other members thus honoured, I am required in this community of scholars to stand and deliver such knowledge and ideas as may be worthy of your attention, or failing that, to own myself unworthy of your confidence.

It is here that I am somewhat at a loss. I remind myself of Satan in the Book of Job when (along with others more worthy) he presented himself before the Lord and was asked, "From whence comest thou?" His answer might well be mine. "From going to and fro in the earth, and walking up and down in it." If I am a scholar, I am a wandering scholar; and I fear that the wandering has left behind more traces than the scholarship! I should, indeed, compare myself not to the itinerant scholar, but to the humble mendicant who went about seeking hospitality—and, not infrequently, singing for his supper or occasionally rewarding his hosts with a traveller's tale. With these I feel close fellowship. I know both their joys and their sorrows.

And, if you will allow me, I should like to retain something of this character tonight. I do not offer you an address fit for your reception in your collective and scholarly capacity. I should like, instead, to speak to you as individuals, as men and women deeply interested, in a very special way, in the well-being of our country and its people. You are concerned, as scholars must be, with the future of a community which you see in length and in depth, knowing its past, and penetrating beneath superficial appearances to the realities of the present. And if what I say should not be unworthy of attention, it will be so because I am at least trying to observe the scholarly method; humility before a difficult and important task; honesty and care in reporting what I know, and only what I know.

May I first touch on the nature of my present work as representative of the Sovereign? Monarchy, I believe, is best described as a kind of society where, by a special personal symbolism, the community seeks to remind itself of its oneness and of its corporate will to see and cherish excellence wherever it may be found. I mention this because, by your name and origin, you are associated in a particular way with such a conception. The original Royal Society was created in order to distinguish those who were devoting themselves disinterestedly to the new scientific learning in seventeenth-century England. They were—to use the old phrase—"divers worthy persons inquisitive into natural philosophy and other parts of human learning." The honour was bestowed by the Monarch, I take it, in recognition of the value to their country of their researches and deliberations.

And yet it is pleasant to remember Pepys's story of King Charles's amusement at some of their early activities. He, we are told, "mightily laughed at" the Fellows of his Royal Society "for spending time only in weighing of ayre, and doing nothing else since they sat." He had, it seems, faith that, given recognition, freedom and encouragement, they would do right, but apparently he did wonder, in a detached fashion, just what he might have started.

He had started something of enduring importance and, despite those qualities which are out of place in learned circles, gave us a great example of how the Crown in so many epochs has been identified with scholarship and learning. Indeed, in the modern age this generally has been regarded as representing one of its obligations. If one could remove from the pages of history the enterprises or achievements in letters and the arts and science which received their first impulse from our Sovereigns, the annals would be greatly impoverished. This Society is but one example of what representatives of the Crown in Canada have been able to do over the years in these fields. You will recall that your Founder himself was, indeed, responsible also for the establishment of the Royal Canadian Academy of Arts, and the National Gallery of Canada.

It is pleasant, I think, to remind ourselves of our relations with the Royal Society of England. Pleasant, too, to remember as we consider the astounding results of the efforts regarded by Charles II with benevolent amusement, that it was our two mother countries which, in the seventeenth and eighteenth centuries, took the lead in intellectual labours which have transformed the world. If your story is blended with that of your elder sister it is indeed one of astonishing achievement.

But I need not say that the Royal Society of Canada has a character of its own, different from that of its counterpart overseas; that it was born in a different age and in a different land. You have your own story of accomplishment and your own special tasks. When this Society was founded, no one in the old world questioned the dignity and greatness of scientific studies, but in this new country everything was to be done. Science and the humanities alike needed nourishing and cherishing. The dignity, the usefulness, the necessity of scholarly work in all its branches, had to be demonstrated in a new and still (in many respects) primitive society.

I wonder whether you are not unique among the learned societies of the world in having in your membership representatives of both the humanities and science. You demonstrate what has been called "the seamless coat of learning" and here biologists, historians, geologists, economists, men of letters, chemists, philosophers, foregather. They are divided in their studies, but united in the intellectual approach which they make to them. Today the work of this Society is enmeshed with that of the national universities, of research institutes, of individual artists and writers, and of the scholarly societies growing every year in numbers and membership, whose annual meetings cluster around this great central gathering, showing what one of your original Fellows foresaw, "la noble contagion d'étude et de travail." Looking back to the humbler beginnings of seventy-five years ago, we can call this a notable, a brilliant development.

By this very fact we can also look for the hidden weaknesses, barely perceptible shadows which the brilliance too easily hides. One of these I will mention, but I will only mention it because I am aware that the active members of the Society understand it much better than I do. It is that your gifted children, each maturing but going his own separate way, may easily forget the value of close and intimate family life in scholarship and the importance of that intelligent and stimulating domestic conversation which it is your special mission to foster and preserve. A Society called "Royal" is, by definition, one and universal. The King's Writ runs everywhere. Historians will recall the tough resistance of the great feudal chiefs long ago to this royal claim. They will remember, too, the danger to the unity and peace of the state when the claim was ignored. May we not discern a parallel in our special situation today? Are we not in some danger of intellectual feudalism?

The power and energy of the departmentalized national societies, the vigour and distinction of the young specialists who form their membership, may draw strength away from the mother society, and may at the same time, separate these strong and healthy offshoots from one another. What can the Royal Society do to preserve the unity of learning which it was intended to safeguard, while encouraging the intense specialization that our knowledge and our needs impose on us? I am interested to learn that your programme for these meetings includes sessions designed to strengthen the community of scholars.

The second source of weakness, I propose to discuss in some detail, because my understanding of it proceeds much more directly from my own observations and reflections.

It is a platitude to say that learning and the arts cannot be supported on strictly economic principles. They must have a patron. Their patron must have enough goodwill to give them support, and enough intelligence to leave them alone. Our attitude to patronage is sometimes distorted. It is easy to assume that, in past eras, the writer, the scholar, or the artist was denied true freedom by his patron, and revealed something servile in his attitude to the prince or the nobleman or the city or the Maecenas of any age who gave him patronage. There were, as we know, sometimes a lack of self-respect on the one hand and unreasonable demands on the other. But we should remember that the old formulae of dedication, which seem to us obsequious, represent merely the forms of politeness of an earlier and politer age. Patrons there had to be, or the arts and science and letters would have withered; patrons there must be today. No matter whose name may stand in the position of honour, the real patron (passive or active) is now the everyday man, the average man. And therefore one must ask whether this collective patron of the twentieth century is any better informed—or even as well informed—as Charles II or others in the past. Certainly anyone can get money today for practical projects; certainly scientific learning never before stood so high in popular esteem; certainly the people of Canada have just given the practitioners of letters and the arts what must seem to them like a gigantic sum. Yet in this Society, we must look beyond superficial appearances. We must ask whether the true nature of the Society, its disinterested concern for knowledge and understanding, really means much more to the community in which we live than "weighing of ayre" did to King Charles.

Obviously, the vast majority of any society must be more or less unaware of the special values for which you stand. I have in mind, however, the critical minority, on whom we must rely to a considerable extent for the maintenance of the standards of civilization. May there not easily develop a dangerous gap between you and this group? May there not be in Canada "two solitudes" represented by the scholarly and learned on the one hand and this minority on the other

—two solitudes perhaps even more to be regretted than those so sharply pictured for us by a member of this Society?

I can best illustrate my meaning by passing on to you some of the questions which run through my mind as I travel through our wealthy, booming, optimistic, enterprising country, with its growing millions—vigorous, able, confident that Canada's century, a little late in starting, is making up for lost time.

As I go about, among those I meet are many of your patrons; not egg-heads, nor long-hairs, but average men and women with an informed interest in your work. These are the people on whose understanding and support you ultimately depend. Are there enough of them? How active are they as your patrons?

Let me begin with one subject on which I have thought and spoken much. How many of us in this prosperous and enlightened country speak or write with purity and precision, with pride in the fact that whichever intellectual "solitude" we inhabit, we have been entrusted with one of the great languages of the world? I am thinking, at the moment, of those whose mother tongue is English and of those whose callings oblige them to take communication—with its possibilities and its pitfalls—seriously. I am not, of course, concerned about those whose energy in the conduct of practical affairs gives their speech a vigour and poetry of its own. I am thinking of the B.A., the M.A., I dare to say the Ph.D., and I certainly cannot except even the thrice-anointed LL.D.

I am reminded by M.A.'s and Ph.D.'s of another matter. I must ask you—for if your judgment fails, no one's can stand—what does education really mean to this important and powerful group? How many are true scholars of the kind that you delight to honour? How many, on the other hand, are diligent, able, even brilliant specialists with no cause to be ashamed, but unfitted—not by their degree of ability, but by the nature of their training—to play a special role in scholarship?

I have another question. How many of us, even those of us engaged in supposedly intellectual pursuits, seek books not as escape, but as food and drink? Here statistics are available. We have good libraries, but too few of them. We have bookshops, but where and how many? People borrow books from libraries; they may buy periodicals for trains and planes, but as a people we have not the habit of buying books to have and to hold, to read and to mark, books where a man has been allowed to set forth his whole thought—not merely fragments tailored to so many thousand words, and then often arranged in a half-tone setting for the striking productions of an advertiser.

And again, how many of us are prepared to support journals which devote themselves to serious and informed discussion of matters which should be of general interest? Ask those unhappy souls who strive to launch and keep them afloat, who—to change the metaphor—set themselves to bridge the gulf between what is scholarly (and presumably "dry") and what is known as "light entertainment."

Here is a further query. How many of us understand and practise something of what a well-known American critic called "poetry as a means of grace"? I am perverting his meaning slightly. He was here recommending it to young clergymen as a necessary refreshment for the soul. But is it not a necessary refreshment for the mind as well as the soul? And is it not meant for scientists and philosophers as well as for clergymen? Surely the practice and use of poetry in a society is an index of imagination, of vigour, of creative power, without which knowledge may be lifeless. Is it not reasonable to say that when a society begins to leave poetry to poets, and to those who must, to quote a dismal phrase, "major in English," its mind is in danger?

Let me return to my M.A.'s and Ph.D.'s, and the meaning of their education. If I had, as it were, to distil its true essence, I would, I think, offer them a generous selection of verse in their mother tongue and ask for their free, untrammelled, uncensored comments. I think that a panel of examiners would be able to select fairly accurately not only the potential professors of literature, but also the original thinkers in chemistry and physics, in politics and industry. I say I think so—I may be wrong. As a Fellow of this Society, albeit merely honorary, how much I would like to initiate this experiment, and I would care no more about the ridicule of the onlookers than the original Fellows cared for the laughter of the Merry Monarch.

I ask these questions very humbly. I do not assume that all the answers will be depressing. In every matter that I have mentioned, I see signs of deep concern among thoughtful people. I see also much to encourage us, but we have no ground for complacency. A retired Canadian businessman, a few years ago, employed a part of his leisure in writing memoirs in which he urged upon his younger colleagues the importance of a sense of urgency. A sense of urgency, I firmly believe, is as necessary in the scholarly as in the business world. The questions I have raised are a matter of concern to the Royal Society. There is, we all know, in the realm of learning today, no ivory tower, no aristocratic preserve, no royal enclosure. In an age of increasing specialization, the Society has come to represent a great chain of communities of scholars. Could not its Fellows, without neglecting their specialties, concern themselves more than ever with the general learning, the whole intellectual life of the community? I say more than ever—I am not impertinent enough to suggest that you have ever forgotten this responsibility. I wish only to convey my personal sense of the need at this time when growing wealth provides so many distractions from the hard, sweet discipline of thought. And may I call it a special need at a time when the increase in public aid demands not less, but much more personal effort from those able to lead and guide our nation in the things of the mind.

THESE SEVENTY-FIVE YEARS

W. A. MACKINTOSH, F.R.S.C.

IL Y A soixante-quinze ans se réunissaient sous le haut patronage du Marquis de Lorne les premiers membres de cette Société. En cette année-là, le vingt-cinq mai, dans cette ville même, dans la salle de la Commission des chemins de fer de la vieille Chambre des Communes eurent lieu les premières séances de cette Société et de ses sections.

Malgré ce début modeste une grande foi animait cette Société. Notre premier Président, Sir William Dawson, dit alors : « We see only the rudiments and the beginnings of things, but if these are healthy and growing, we should regard them with hope, should cherish and nurture them as the germs of greater things in the future. »

C'est ainsi que s'exprima l'éminent géologue. Il était bon aussi que le distingué vice-président, l'honorable Dr. Chauveau, rappelât : « Il y a longtemps, bien longtemps, que l'on fait de nobles efforts pour la culture de l'esprit humain, sur les rives du Saint Laurent. »

Depuis soixante-quinze ans bien des choses ont été accomplies dans le domaine des lettres et des sciences dans ce pays. Nul d'entre nous n'oserait prétendre que la Société Royale ait été la seule influence ou même l'influence dominante dans le développement de notre culture nationale. D'autre part, on ne saurait surestimer l'importance de ces réunions d'érudits et de savants pour l'établissement d'objectifs élevés et pour l'encouragement apporté à ceux qui, sans elle, travailleraient dans l'isolement.

Ce n'est pas le moindre titre de gloire de cette Société que d'avoir assuré une étroite collaboration fondée sur le respect mutuel et sur une compréhension sans cesse plus profonde dans tous les domaines des belles-lettres, des sciences et de l'érudition, entre ceux qui pensent et s'expriment en français et ceux qui pensent et s'expriment en anglais.

Notre collaboration devient de plus en plus confiante à mesure que nous reconnaissons mieux que la culture s'élève au-dessus des frontières nationales et raciales tout en pénétrant plus profondément par ses racines dans notre patrimoine canadien.

The greater part of the programme of the 1957 meetings of the Royal Society of Canada is devoted to "Our Debt to the Future." We look before and after at the interim and final reports of the Royal Commision on Canada's Economic Prospects. With such notable groups of Canadians poised, each with hand raised to shade his eyes, straining like Lief Ericson to discern what breaks the horizon of the future, the President is left free to turn his attention to the past. He does this not with any intention of recalling the triumphs and defeats of this society but in the hope of gaining some deeper insight into present and future. All history, said Croce, is contemporaneous. The trick of the historian is to discern what of the past is relevant to the present and not to be misled into accepting a reflection of the present for history.

My way has been made easy by the government which, either because it recognized time as a dimension or by inadvertence, has made such temporal arrangements that as I speak the turning earth draws a curtain of silence from east to west over the contentious issues of the day. None but the most perceptive and least impetuous of the British Columbians could be influenced in his electoral choice by anything which I say. These fortunate circumstances relieve me from genuine anxiety. I might, for example, have quoted, in innocent piety, a sentence from the first President's address—"A locust, a midge, or a parasitic fungus may suddenly reduce to naught the calculations of a finance minister." How could I have convinced my hearers a few hours ago that I was not, by this indirect means, referring to the much discussed contemporary phenomenon of the federal surplus repeatedly outguessing successive finance ministers? How could I have been sure that some inflamed supporter or rabid opponent of the government of the day did not find in this innocent sentence of a guileless scientist scurrilous references to our three loyal oppositions? But I am freed from all such fears.*

The Royal Society of Canada is a society of persons and was founded by persons. It is clear that without the initiative and vision of the Marquis of Lorne the society would not have been established. No doubt, on his part it was something of a missionary effort in a pioneer country, and perhaps as a Campbell he was concerned for the future of a country under the government of a Macdonald. Yet we have reason to be grateful to him for the achievement and the precedent. We have reason also to be grateful to such successors as Grey and Tweedsmuir who made powerful contributions to the intellectual life of this country. We have cause for very special gratitude to our Honorary Patron, His Excellency,

*This address was given in the evening of the day of the federal election of 1957.

our present Governor-General, who has worked so assiduously and with such perception for the strengthening of the intellectual and cultural life of this country and who has the satisfaction of seeing a growing number of the recommendations of the Commission over which he presided put into effect and bear fruit.

But the Marquis of Lorne did not work with unlettered pioneers. As Lawrence Burpee wrote twenty-five years ago:

One need only think of what Sir William Dawson meant to McGill, Sir Daniel Wilson to Toronto, and George M. Grant to Queen's, of what the Geological Survey owes to George Dawson, the Meteorological Service to Charles Carpmael and the Dominion Experimental Farms to Saunders, of Osler's achievements in medicine, Macoun's in botany, and Sandford Fleming's in engineering, and of the contributions to Canadian literature of the poets, Frechette and LeMay, the historians Sulte and Verreau, and the novelists Kirby and Marmette, to realize that the Royal Society counted among its charter members a number of men of unusual ability and force of character.

Neither did the first Patron of this Society work outside the circumstances of his time. In Britain the Victorian era was in the late harvest. George Eliot and Clerk Maxwell had died recently. The lives of Carlyle, Darwin and Disraeli were to end in 1881 and 1882. Browning, Matthew Arnold, Tennyson were living figures even though their work was nearly done. Gladstone had still to re-emerge for his final act. Sir John Clapham's "age of free trade and steel" was yielding its fruits.

The United States had turned its back on the civil war and was concerned with successive transcontinental railways, the agrarian revolt, free silver and the integration of Lake Superior iron and Pennsylvania coal. New England glowed mildly in its "Indian Summer." Parkman had already published his *Frontenac, The Old Régime* and *Lasalle.*

In Canada, the project of Confederation was incomplete. The political structure had been breached by Mowat and in great measure transformed by the decisions of the Judicial Committee of the Privy Council. The West had been acquired but was still empty, as Edward Blake monotonously repeated. More than 76 per cent of the population was in Ontario and Quebec: only 4 per cent was west of Ontario. The National Policy had been launched but Tupper did not complete it till 1887. Confederation was a daring, indeed desperate, project of nation building with most unlikely and ill-matched materials. In 1882, even the beginning of success was in the future.

And yet now and then there was a promise of sunshine. The bold project of the Canadian Pacific Railway was put together in 1880 and, after a hesitating start, was executed with great vigour. It was made possible by, and in some measure created, a brief, buoyant wave of expansion. Canadian exports reached their peak for the decade in 1882. Imports reached figures in 1882–3 which they did not again achieve till 1898. Immigration which had been 26,000 in 1876 rose to 112,000 in 1882 and 134,000 in 1883, figures which might command respect even in the present decade.

With the railway in prospect, the Homestead Act began to function. Homestead entries in the West rose from a pitiful 350 in 1876 to a sanguine 7,000 in 1882. Winnipeg opened its first grain exchange in 1881 and in 1883 there was a record export of more than 8,000,000 bushels of wheat and flour. George Stephen had brought in those great captains of men and mules, Van Horne and Shaughnessy, to drive ahead the lagging construction of the Canadian Pacific. Something of what John Macoun and G. M. Grant had predicted seemed to be coming true west of the Great Lakes.

And yet it was a false start. Almost immediately the price of wheat fell disastrously. After the first promising crops came the three destroyers —frost, the grasshopper and drought. The first Winnipeg Grain Exchange closed its doors. In 1889 Canada consumed more wheat than she produced. The Northern Pacific, in some degree the prototype for the Canadian Pacific, went bankrupt in 1883. In the United States and England, the years from 1884 to 1896 were labelled "depression" or "deep depression" except for brief and feverish revivals around 1890. It may well be that had the Royal Society of Canada not been planned and launched in 1881 and 1882, it might have waited another fifteen or twenty years to appear.

These seventy-five years between the founding of the Royal Society of Canada and the setting up of the Canada Council have been years of great but uneven growth—years bristling with political and economic difficulties. Now that every Canadian can tell glibly just what our population and national product will be in 1980, there is some profit in looking back at some of the obstacles which have been surmounted.

Sir Wilfrid Laurier, and several others, are reported to have said that Canada was a difficult country to govern and it is easy to understand the difficulties inherent in our distances, our two languages and our diverse origins. (It is well to remember, however, that just as a man about town may eat out on a slender stock of other people's stories, many a Canadian politician has made a career of other people's prejudices and misconceptions.) Sir John Macdonald might well have said that Canada was a difficult country to put together. It was not clear when he died, 288 years after the first settlement, that what he had put together could live and develop into a national unit.

Some obvious divisive forces, of climate, distance, language and religion, are important but not decisive. A major geographic and economic obstacle of historic importance has been the lack of adequate hinterlands for the St. Lawrence and for the Atlantic ports. The Atlantic harbours have been disproportionately large and commodious for any hinterland which could be made tributary to them despite railway building, freight subsidies and preferential duties for direct Commonwealth shipments. Throughout much of our history, the noble St. Lawrence has been a rather pathetic river in search of a hinterland.

The development of Canada, as of the rest of North America, was an extension of the trade area of the European metropolises. In this the St. Lawrence, the conspicuous eastern gateway, seemed to be of central importance. Indeed, without the St. Lawrence and some vision of its continental role, there could have been no Canada, or to put it more cautiously, it would have been a totally different and probably never an independent country. It was Lasalle and Frontenac who, turning their backs on the sea and its fish, saw North America as a continent with its greatest resources in the interior and the St. Lawrence as an instrument of imperial strategy. They saw too that the most accessible and valuable hinterland of this river and its lakes lay to the south in the area between the Ohio and the Mississippi.

This was the territory which our French-speaking forerunners were unable to withhold from the British armies based on the American colonies. This was the area which the British imperial power retained in the province of Quebec in 1774, only to relinquish it in 1783. This was the area which, as Professor Creighton has told so well, the Montreal merchants tried desperately to bind economically to the St. Lawrence in a long period which finally came to an inglorious full stop when the Grand Trunk rolled ineffectively and unnoticed into Chicago in 1880.

Lasalle's project of empire, an attempt to control the interior through control of the river, was almost certainly foredoomed to failure. The St. Lawrence valley itself could not generate and maintain the necessary weight of population to control the interior. Had the St. Lawrence remained the only traffic outlet from the Great Lakes, it would at some juncture have been taken over by the United States just as surely as Andrew Jackson defeated John Quincy Adams for the Presidency. What relieved the pressure and diverted the interest of the United States before the weight of her population became overpowering was the building of the Erie Canal, linking the Great Lakes to a warm-water port, avoiding

Niagara and the St. Lawrence Rapids. In 1774, the Canadians held the river, nearly the whole of the Great Lakes and the essential hinterland. In 1825, the United States held the hinterland and was populating it at an awesome rate. It had joint ownership of the Great Lakes. It had provided an eastern gateway which, at the time, had decided advantages over the St. Lawrence. Small wonder that the United States was less interested in Canada than in the territories beyond the Mississippi.

The area between the Ohio and the Mississippi was the Old Northwest territory. The settlement of Tennessee and Kentucky and the Old Southwest had preceded it, but Ohio became a state in 1803, Indiana in 1816, Illinois in 1818 and Michigan in 1837. Here was the area of agricultural land which Canada lacked, suited to the technical knowledge of the day, ready to yield its wealth to the axe, the plough and the simple household arts. It quickly supported a weight of population which gave point to Henry Clay's "American System" by providing a market for manufactures, generating demands for transportation facilities and ultimately tipping the balance in the Civil War. If the existence of this area was important in the history of the United States, the lack of it was crucial in Canadian development. The territory north of the St. Lawrence and the Great Lakes, narrowed by climate and reduced drastically by the Canadian Shield, offered no comparable pioneer farm area and few other resources for which there was market need or knowledge to exploit.

What may seem to you a sketchy digression into the history of the United States is not as irrelevant as it would appear. There was not in 1882, despite the new Royal Society of Canada, a viable Canadian economy capable of a national integration and differentiation of function. If I may simplify the economic facts, while not denying the influence of manifold other forces, there was a fatal lack of adequate agricultural and other resources suited to the markets and technical knowledge of the day. There was also lacking the technical knowledge to create markets and to convert useless space into valuable resources. The territory between the Ohio and the Mississippi was rich in the resources amenable to contemporary skills and knowledge, and was capable of supporting decisive densities of population—arable land in large continuous areas, coal, iron and oil. In 1882, Canada had uncertainly in hand a great nation-building project which had as yet yielded only discouraging results and whose success in the future was highly uncertain. Looked at from one point of view the project was to find and develop within Canadian boundaries resources equivalent to those lost in 1783 or to use different words it was to find an adequate hinterland for the St. Lawrence.

The settlement of the Prairie Provinces was made possible by such Canadian achievements as the building of the Canadian Pacific and the discovery and development of early maturing wheats. It was made feasible by external shifts in prices and costs which reflected the greater economy of overseas cereals in an industrialized Britain. It needed also a host of innovations borrowed chiefly from the United States. This settlement was the first rapid and massive movement toward a Canadian economy. In twenty years the population of Canada increased by nearly two-thirds. Improved farm land increased one and one-third times. The St. Lawrence added a huge farming area to its hinterland and acquired its first massive Canadian traffic other than timber. In 1889 Canada consumed more wheat than she produced. When settlement was complete she was supplying 40 per cent of the world's wheat exports.

After the conquest of wheat came the conquest of newsprint and power. The growth of the metropolitan press and the inadequacy of United States sources of power and soft wood produced the unlikely phenomenon of a Republican Congress removing the duty on newsprint in 1911. Climate, glacier and beaver had combined to produce spruce and water storage in hitherto undeveloped areas of the Canadian Shield. Improvements in the generation and transmission of electricity gave further advantages to water power in an industry which is lavish in its use of mechanical power per worker. In twenty years the industry became our second major exporter. While not confined to the St. Lawrence Valley, it gave the St. Lawrence its second great hinterland.

The gold-mining industry for a time contributed the third of the great exports and deepened the hinterland of the St. Lawrence. Its rise and expansion were, however, to a degree fortuitous both in discovery and in the revaluation of gold in 1933.

The rapid expansion of base-metal production which began in the 1920's again followed a familiar pattern. The decline of competitive sources of supply and the invention of new technical means of recovering metal from complex ores were the roots of its growth. Increasingly, markets and knowledge, time and applied science, furnished the basis for investment.

It has been convenient to relate developments to the historical setting of the St. Lawrence Valley and they still have great relevance. But the simplicity of that pattern had long since been broken. The availability of the Panama Canal by 1920 re-oriented or more correctly re-occidented British Columbia. Her coastal resources, accessible to Europe and the Atlantic coast, became more valuable than her inland resources marketable in the Prairie Provinces. The greater importance of her market gave direct stimulus to other regions.

In the years since World War II there has been an astonishing increase in the speed and mass of development but no conspicuous change in pattern. Despite tremendous changes, we are still developing, as we were at the beginning of the century, great bulk exports albeit with an increasing amount of processing and refinement. We are still dependent on markets abroad. We are still the field for investment of great sums of external capital.

The great expansion of this country in the past twelve years, whether we look on it with admiration or dismay, has been stimulated and conditioned by a number of circumstances which have historical significance. I shall not seek to explain or account for the great rise in Canadian and western population which has taken place in utter disrespect of the once-hallowed logistic curve and to the humiliation of demographers. I shall merely list briefly some of the circumstances of economic growth which seem to me to be relevant and illuminating. First, of course, the low capital investment of the preceding fifteen years of depression and war had built up a deficit which was waiting to be filled. Second, the enormous increase in the material intake of modern (particularly United States) industry produced immediate and prospective markets for materials and conspicuously for metals. For these markets, the customary but declining sources could not long be adequate. Third, technical advances in the means of discovering, mining, extracting and treating metallic ores made possible the discovery of unsuspected resources and the use of hitherto unusable resources.

Just as Canada, having built a transcontinental railway, had to wait for favourable markets, for early-maturing wheat, and for improved methods before she achieved success in establishing prairie agriculture, so we had to wait until markets and methods had caught up with our other resources. True, there has been some good luck, but it has been rather in the extent of the resources than in their discovery.

The fact that our recently tapped resources of iron, nickel, copper, uranium, oil, natural gas and water power are mainly in unpopulated or thinly populated parts of this country has historic significance. The hinterlands of the St. Lawrence system and of Canadian metropolises have been widened and deepened. There is a kind of historic justice in the fact that between 1882 and 1957 we have found within our own boundaries, the equivalent of the rich farm land, the iron and the oil which lay in the triangle between the Ohio and the Mississippi. We have found also an abundance of uranium which may remove the penalty of our ill-located coal resources. The Canadian economy, which appeared viable in 1882 only to the most stubbornly optimistic, has in 1957 an assured life though one which we may be certain will not be without its

vicissitudes. It is significant also that once more the St. Lawrence is one
of the centres of attention, not merely because its new hinterlands
demand it as an outlet and for the interchange of their products but
because the old hinterland requires the import of materials to supplement
its waning resources and power to serve its enlarged needs for energy.

But there are matters of even greater import than these material
advances. Canada is in a new era not only because the Royal Society of
Canada has been in existence for seventy-five years, nor yet because the
Canada Council has been established, not because the Royal Commission
on Canada's Economic Prospects has produced exhilarating statistics as
difficult to believe as a compound interest table, and not alone because
the St. Lawrence is about to resume its historic Canadian role.

We are in a new era because we are as a people meeting challenges
which we have not met in such force before or the nature of which we
are now better able to recognize. Our life is no longer shaped by the
deficiencies of the past but by the opportunities of the future. But unless
we have learned from the past, we shall not understand the opportunities.

Let me mention, as illustrations, two or three areas of significance and
concern. The development of the northern half of this continent has been
delayed so long because it required a level of industrial and urban
civilization, a degree of technical knowledge, a massing of capital which
did not earlier exist. It is not so long ago, though in another age, that the
late Griffith Taylor envisaged and even advocated the northern extension
of settlement by a tough peasantry asking nothing but space, adequate
rainfall and freedom to multiply. Nothing could be clearer in the light of
today's experience than that the development of the north will proceed,
if it does proceed, not by peasant shrewdness and endurance but by
accurate technical knowledge and skill, by the products of applied
science, the aeroplane, radar, the magnetometer, the tractor. The
exploitation of its resources will require massive investment, highly
technical equipment, and skilled application and direction. Canada
requires for its development not a lower but a higher intellectual capacity
than most others. What we face is not material ease but intellectual and
moral challenge.

The necessity of scientific training and capacity are at the moment
obvious. Less obvious but equally necessary are other types of learning
and skill. Canada will continue to be a difficult country to govern, but
the chief difficulties of the future will not spring, I suspect, from our
differing cultural, religious and linguistic origins. It is evident that this
country will embrace vastly greater metropolitan centres with intractable
problems of physical organization and political accommodation. It is

equally evident that we shall have to the north a perpetual frontier with scattered and shifting concentrations of population. Metropolis and frontier will be linked by many cultural, political and religious ties but also by powerful corporate and union organizations. It is going to take informed understanding and real knowledge of social, economic and political problems to combine freedom, order and change in such a country. The slogans and the appeals to the prejudices of the past will have lost their relevance. The translation of the commonplace into the jargon of the seminar room will not meet the requirements.

Let me mention another area of concern. At times in our history we have been nervously concerned with the attitudes and intentions of the United States alternately seeing a pattern to be followed or a fate to be shunned. As for political absorption, the fact is that seldom if ever did we offer any serious temptation. With our growing population and national integration and in today's climate of opinion, no one but an occasional angry congressman thinks of it.

There is at present in this country a widespread alertness to the dangers of economic and cultural absorption. Certainly United States capital has acquired controlling ownership in wide sectors of our industry and natural resources. United States union organization has achieved varying degrees of dominance in Canadian industry. Ownership is, of course, exercised and operations carried on within Canadian laws. And yet there is an area here for alertness though not for alarm.

In organization and practice, United States corporations in Canada fall roughly into two classes. There are those which in their Canadian subsidiaries offer minority stockholdings for Canadian purchase; establish Canadian boards of directors which are more than advisory boards; offer opportunities to Canadian executives; develop research centres in this country and participate as Canadian companies in the pattern of Canadian life. In contrast, there are others which, though dominant in their industry, operate branch plants or offices in Canada in the same way in which they would operate a branch in Wichita or Oshkosh. Decisions and research are localized in United States centres. Policies which may be relevant enough in the United States transfer to Canada attitudes toward government, education and community life which are foreign to Canadian thinking. It seems probable to me that it was the missionary ancestors of the directors of these corporations who put the island natives into Mother Hubbard dresses. There is in this country a strong tide of informed judgment which favours the first type of corporate organization and policy and is increasingly resistant to the second.

It is also characteristic of this present era that there is a widespread, though uneven, alertness to the threats to, or the lack of, Canadian culture. A national culture is not a direct object of endeavour. It is not created as a gown by a designer. It is a by-product. Further, a country can have a truly national culture, incredibly bad. Canadians should aim at what is excellent intellectually, aesthetically, socially. If it is real, it will ultimately prove to be Canadian but its justification will be that it is excellent. The destroying sin is imitation. Writing or painting or music is no better because it is Canadian. Its virtue must be that it has reality and integrity and gives insight into great experience. Much of that experience will be recognizably Canadian, because we now have a Canadian experience which is coherent. If we give opportunity and encouragement to what is first-rate in our own experience, achievement and talent, we shall be protected against second-rate borrowings from larger countries.

The seventy-fifth anniversary of the founding of the Royal Society of Canada, the institution of the Canada Council, the ninetieth anniversary of Confederation, the reconstruction of the St. Lawrence Waterway, are all interwoven strands in the emerging pattern of a Canadian economy and nationality. These seventy-five years have proved and made greater the vital national importance of all that a society of letters, philosophy and science represents in Canadian life.

THE ROLES OF THE SCIENTIST AND THE
SCHOLAR IN CANADA'S FUTURE

W. A. MACKINTOSH, F.R.S.C., *Chairman*
DAVID L. THOMSON, F.R.S.C.

W. A. MACKINTOSH

I HAVE BEEN GIVEN THE HONOUR of opening the symposium which has been arranged as the main feature of our programme on this, our seventy-fifth anniversary. This symposium, in its several parts, comes under the general title "Our Debt to the Future."

This Society was established seventy-five years ago, having for its object the promotion of literature and science within the Dominion. Having recently discarded the Dominion, I take it our efforts are now concentrated on preserving and promoting literature and the sciences and that our conviction is that it is on these that our future largely depends.

The symposium and its objects will, I anticipate, unfold as it proceeds and those who participate will disclose the direction and limits which have been set. Any introduction which I might give would run the serious risk of misrepresenting the programme committee and misleading the audience. I am left with the unnecessary, less risky and more agreeable function of introducing the speaker who, in opening this symposium, will examine the role of the scientist and the scholar in Canada's future. Dr. David Thomson, Vice-Principal of McGill University, Fellow of this Society, is one of the most distinguished of our scientists and among the most articulate of our scholars.

DAVID L. THOMSON

MY TASK, as I understand it, is somehow to prepare you, somehow to make you receptive to the wiser words which my learned colleagues will pronounce at our later sessions. Clearly, I must not say what they are going to say; that would be to steal their thunder. Nor may I say anything to contradict them; that would confuse you, annoy them, and humiliate me. It is a task whose delicacy is measurably increased by the fact that I have not been given the slightest indication what they *are* going to say.

I think that I must also try to steer a middle course between two extreme views of Canada's future. There would be little point in going on, if I sincerely believed that in a year or two we should all have been hydrogenated and lithiated and strontiated out of existence, that *Homo sapiens* will be homogenized. We do well to "be afraid of that which is high" and to "fearfully o'ertrip" the DEW-line; but our dread must not harden into despair. Let our world not end with *both* a bang and a whimper. Nor would it be helpful, I believe, if I were to accept a picture of our future so rosy that no problems would remain: a science-fiction Land of Cockayne, in which every automobile would be equipped with a portable parking-space; every bank would have, as well as the slot in which you can deposit money at night, a slot into which you could reach for a large handful of bills whenever you passed by; and both the Brooklyn Dodgers and the Metropolitan Opera would have moved to Aklavik to secure larger, wealthier, and more knowledgeable audience. Let me proceed, then, within these self-imposed limits.

It is said that at the moment of the expulsion from the Garden of Eden, Adam turned to Eve and said, "We live in times of transition." This observation, however trite, must be made again today. In a generation we have seen Canada transformed from a nation primarily agricultural into a great source of mineral wealth and a significant industrial power. Similar transitions have taken place elsewhere, and are occurring elsewhere today, and the end is not yet. But in Canada the change has been so rapid and so sweeping that many refer to it as a revolution.

In a revolution it is extremely difficult to know what is really going on. It seems to me that most revolutions are directed against institutions which are already crumbling to dust, and that their outcome is often very far from that anticipated or desired by either side. The conspiracy against Julius Caesar led directly to the establishment of a long line of autocrats. The Arab civil war of 750 moved the Caliphate from Damascus to Baghdad, not back to Medina. Wat Tyler's Revolt of 1381 seems to have done nothing to hasten the already evident decay of serfdom. The American Revolution might be viewed as an outbreak against the system of having Ministers not responsible to Parliament, though it was Britain rather than America that learned this lesson. The French Revolution was an attack on the landed nobility, whose estates were already sapped by the King's deliberate concentration upon the central Court, and it did nothing to protect the proletariat against the rising menace of the unscrupulous *entrepreneur*. The Russian Revolution established a machinery of state far more comprehensive and repressive than that of the Tzar.

In such terms, then, let us look at peaceful revolution in Canada in our time. In the thirties, the wheat economy proved itself inadequate to sustain a population large enough to justify possession of Canada's vast area; and change was bound to come. We were fortunate in that the discovery of, and the demand for, petroleum in the west, uranium in the north, and iron in the northeast, coupled with the wartime stimulus to production and manufacture, made it possible for the change to be both peaceful and profitable. But since the secular trend is to increase the quantity of product per man-hour—a trend which we dramatize in industry by the term "automation," but which is also very evident in agriculture—we may not yet have solved our problem of sustaining a population increasing fast enough to justify our retention of this land-mass. No doubt new sources of power will be tapped and new mineral wealth uncovered. I remember once saying to the Premier of Newfoundland that I thought it likely that some such discoveries would be made in his province; Mr. Smallwood looked at me exactly as if I had pronounced myself in favour of the theory that the sun would rise tomorrow; what I thought probable, he took for granted; and I dare say he was right. The nearer future is bright enough. But a time will come when the stage of exploration is over and the costs of further development become frighteningly high; and a time will come when efficient industry is *not* concentrated in the grasp of a handful of nations. Since men must eat, our descendants of that distant day may be glad to re-discover Canada's position as a great agricultural nation. Many of the great industrial powers do not have this strentgth in reserve.

Let me state the case in another way, looking not quite so far ahead. We hope to maintain a high standard of living; which means that we must compcte successfully in the markets of the world. We cannot do this on the basis of having a virtual monopoly for a few products; nor on the basis that we *at this moment* possess certain skills that millions of our fellow-men do not; they will not be slow to learn. We can do it only if our products are better, or our production more efficient—only by a continuous process of improvement. If we are like Alice, if it takes all the running we can do to stay in the same place, we shall be left behind, we shall have failed. We must stay with the leaders: and here is the point—if we succeed, it can only be through the successful efforts of our scientists and technologists.

I need not labour this point. Yet I should like to read into the record a series of helpful definitions which Dr. Steacie evolved for the Universities' Conference in Ottawa here last fall. The ideas are his, if the actual words are mine.

The "pure scientist" is he who bends his eye upon Nature and asks himself, "What is this, and why does it behave in this way, and how?" He asks himself, for example, as Sir Alexander Fleming did, "Why do bacteria fail to grow in the neighbourhood of colonies of *Penicillium notatum?*" The "applied scientist" is he who picks up the answer and asks himself, "Can this discovery be applied to human welfare? Will penicillin, for instance, exert its bacteriostatic action within the human body? Is it sufficiently safe for use under such conditions?" If the answer be "yes," then comes the turn of the "technologist," who asks himself, "How can we produce penicillin in quantities sufficient to make any impact upon the general welfare of the human race?" So the plant is built, and is operated by the "technician," who need ask himself no questions more searching than "Which button do I press when the needle on this gauge passes this mark?" One point Dr. Steacie did not make is that any one man in his life may play two or even more of these roles. A point that he did make, and that I wish to emphasize, is that the successful performance of each of these roles is essential to the continued prosperity of our country. It matters not that none of them may think of their task in such terms: that they say rather "let us run with patience the race that is set before us." Things become more difficult when they also say, quoting from the same verse, "Let us lay aside every weight." But I have already trespassed on too many fields to venture upon the dusty answers of "Industrelations." I shall merely remind you of one problem we are all aware of: how shall we siphon off an aliquot to be the teachers of the next generation? Where are our seed-potatoes?

The organizers of this symposium, in their wisdom, have seen fit to make a distinction between the "scholar" and the "scientist." I go along with this, with reluctant loyalty. In one sense at least the implied distinction is valid; unfortunately, many, but not all, scientists are deficient in an ability possessed by many, but not all, scholars: the ability to say what they mean in perspicuous and engaging English. This is particularly evident in the Ph.D. dissertation, not because there is much improvement in later life, but because the scientist's later writings are likely to be cast in the telegraphic jargon on which most editors of scientific journals insist: I invite those unfamiliar with its conventions to glance at the *Journal of the American Chemical Society*, in which it is the rule that the name of a complex substance must not be spelled out twice in a single article—much as it is a rule in heraldry that the name of a colour must not be repeated in the blazoning of an escutcheon. It is a curious, contorted, crippled style, not easily learned. "The vocabulary of Bradshaw's time-table," said Sherlock Holmes, "is nervous and terse, but limited." This is true of many journals in the natural sciences; in the social sciences, on the other hand, the vocabulary often strikes me as enervating and *not* terse; but still limited by its preposterous predilection for polysyllabic periphrasis.

I return to the limping linguistics of the typical graduate student in the natural sciences. My colleagues in modern languages frequently report that Mr. So-and-So, Ph.D. candidate in chemophysiobiology, has done a miserable job in his required translation from French or German; but that we must not be too hard upon him, since it is clear that the English into which he is translating is not his native tongue. They apparently do not note that his name is Smith or Macdonald, and they are unaware that he has never quitted this continent. The man simply cannot write English.

The explanation is simple and the cure obvious. To the undergraduate in arts are assigned readings, quite often in books that have some literary flavour; he is required to write essays on his readings, and his efforts are returned to him with vituperative comment. He improves. But if the undergraduate in science is assigned readings, these are not calculated to better his style, and essays are rarely required of him. His ability to write does not improve. The arts department thinks of its course as lectures, reading, and essays. The science department thinks of its course as lectures—and laboratory. The laboratory work is rightly thought of as important, but wrongly thought of as cancelling the necessity for reading and writing (surely nobody will pretend that the laboratory note-book is *writing*?), and the administration sees no reason for providing one department with both demonstrators and essay-readers—if, indeed, essay-readers with any standards could be found.

My last phrase adumbrates the weakness of our system of language tests for the Ph.D. candidate. They are of no value unless the professor himself can and does read foreign languages and expects his graduate students to do so too; being himself probably a product of our system of intensive cramming followed by swift forgetting, the chances are that he does not. The point was well made by Dean Dansereau in the *Studia Varia* recently issued by this Society. I myself, as a graduate student, worked under one Nobel prizewinner who knew no French and another who knew no English; they both regretted it: said one, *"Es ist ein Jammer, wenn man kein Englisch spricht."*

You will think that I am dodging one portion of my assignment: the role of the scholar in Canada's future as distinct from that of the scientist or technologist. I have been skirting it, but I have not faced it head-on. To do this, I have to paraphrase what I have said before, in an article which (fortunately) few of you have read. We are told that industry and government require recruits who can think and speak and write. *Benissimo!* But if those trained in science and engineering cannot do these things, that is the fault of the way in which we teach science and engineering, not of these subjects in themselves. We are told that we need something to offset the increasingly mechanical and materialistic character of our society, and to form and maintain standards of good taste. "Where there is no vision the people perish," but what vision do our people have? Television! Certainly it is frightening to think that swing-and-sway and rock-and-roll must inevitably develop into jive-and-drive and love-and-shove, and the Bikini bathing suit irresistibly evolve into the Bitwini. Yet we my legitimately wonder whether such currents can be stayed by the possibility that one per cent of our population may prefer Chaucer to Cadillacs and Dante to Daiquiris. It is my conviction that the long-playing record has done more to humanize the Philistines than a battalion of professors of philosophy.

The humanities "of course" teach values, and the sciences "of course" do not: unless one happens to think that the selfless search for truth, even it be unpalatable, has "value." A friend of mine wrote his Ph.D. dissertation on the occurrence of the accusative-infinitive construction in Livy; that he became a humane and delightful scholar seems to me to be purely coincidental.

Yet my complaint of the humanities is, not so much that they make illegitimate claims, as that they often fail to make the legitimate ones. A language department, for example, commands two mighty educational tools: in translation, it can insist upon the sharpest accuracy in the use of words, as precise as arithmetic and more difficult; and in literary criticism it can challenge the student with ideas for which it is hard to find words at all. More generally, we may say this: to be learned in a science, it is necessary and *almost* sufficient to know what the scientists have said; to be learned in a literature, it is necessary and *almost* sufficient to know what the great authors wrote. The professor and his textbook here come between the student and his true subject like a semi-transparent screen. It is precisely here that the great educational possibilities lie! For the student can pierce the screen and get to the authors himself and form his own conclusions, half-baked as they may be; and then, returning, he compares his conclusions with his professor's, and so gradually, even despite himself, he comes slowly to recognize the difference between his own hasty gulping and rejection, and the informed and subtle acumen that is the heart of true scholarship.

The oncoming "crisis in education" has been too much discussed for a contribution from me to have any usefulness; except that I might raise the question, what patterns do we expect our overloaded universities to follow, and with what objectives? I can make this question more crucial by describing for you three extreme types as we see them among the institutions of higher learning in the United States.

The first is the typical four-year "liberal arts college," co-educational or not, situated near some small country town but not especially serving the local area. Its chief aim, not unlike that of the English public school, is to "build character," and to that end it encourages extra-curricular activities, formal and informal, and it vigorously encourages its staff to participate in them; so that when the student graduates he knows and is known by most members of the faculty, not only his own teachers. There is, indeed, little else for the faculty to do! Facilities for research are meagre, and the academic community is a microcosm having few external contacts. The curriculum makes a virtue of emphasizing breadth rather than depth; necessarily, since the college is unlikely to attract or retain scholars capable of teaching at advanced levels.

The second of my three extreme types is the municipal college or university. In its early stages at least, it resembles the first in that again its curriculum emphasizes breadth rather than depth, and for the same fundamental reason; but the curricula will tend to be less "liberal" and more "vocational" in emphasis. In all other respects the city college is very different from its country cousin. It is dedicated implicitly to the proposition that education is a matter of acquiring the necessary total of "credits" by examinations (a form of academic bingo), and its aim is to make this possible at the lowest cost to the student. Student and teacher alike are commonly employed elsewhere, so that there is little scope for extra-curricular activities, and, of course, none for research.

The apex of my triangle is occupied by those few great universities that are so wealthily endowed that they can not only maintain a high ratio of staff to students (so that small classes are compatible with light teaching loads), but can afford to attract to the staff scholars of international reputation. This requires not only generous salaries but also the provision of ample facilities and opportunities for research, and the seniors at least have a good deal of liberty to decide the amounts and the levels of the teaching they will undertake. So, too, the student has a good deal of liberty to decide how vigorously he will avail himself of the incomparable educational opportunities this system can provide at the higher levels.

These, then, are the three vertices of this triangle; with the great majority of universities occupying some more central position, subject to pressures in all three directions and perpetually forced to compromise. A university worthy of the title must have some great names on its roster (by this it is judged by its peers), and this, in turn, demands a climate not inimical to research—which is expensive, especially in terms of time; research is not done in an hour between lectures. The struggle to make ends meet often leads to toleration of undesirably large classes. Perhaps the greatest danger is that the staff will become divided into researchers, who look well in the shop-window but virtually do not teach, and teachers, refugees from the colleges, who have neither interest in nor capacity for such research as they may mechanically perform.

It seems to me important, that in planning our educational future, we should have these various patterns and various objectives in mind; we shall then understand better the natures and the consequences of the pressures to which we shall be increasingly subjected.

There is one word in the title assigned to me to which I have, so far, given little explicit attention: the word "Canada." A witty student skit observed very recently that Canadians spend half of their time explaining to Americans that they are unlike the British, and the other half explaining to the British that they are unlike the Americans. It is incontestable that much of our thinking is coloured by our awareness of the size, wealth, and power of our neighbour to the south. In many ways she shocks and alarms us. We may feel that the American male has mis-read his historic documents and believes himself entitled to love, libertinism, and the happiness of pursuit. We may think that the road down which she hastens so eagerly is but a *Via dollar-osa*. The incalculable oscillations of her foreign policy may lead us to suspect that Washington, D.C., has been re-wired as Washington, A.C. But she is as inescapably there as the weather; we may chafe at times, but we must accept the fact. Mr. C. D. Howe has said that Canadian nationalism must not be merely negative, not merely critical and rejecting. We have elements here from which we could synthesize something positive, something different and our own, a new and harmonious chord in the symphony of the New World.

I do not think, however, that we shall achieve this unless we can find some way (and I do not know where it is to be found) of stimulating the spirit of enterprise and adventure in more of our young people. Too many of them, I feel, are satisfied with mere competence, worthy but dull, drearily conformist. When we cry "Expansion!" they ask about pensions; they worry over sickness-insurance when they should be insuring against prosaicness. Of the scientists and students working in our far north, for example, far too few are Canadian-born.

Let me sum up all I have tried to say by quoting a sentence spoken last year by that master phrase-maker, Sir Winston Churchill: "Man does not live by bread alone. Our aim must be men and women not only skilled and diligent in earning a living, but having access to those deep wells in which lie the secrets of the future."

THE PENALTIES OF IGNORANCE OF MAN'S BIOLOGICAL DEPENDENCE

E. G. D. MURRAY, F.R.S.C., *Chairman*

K. W. NEATBY, F.R.S.C.—IAN McT. COWAN, F.R.S.C.

G. H. ETTINGER, F.R.S.C.—R. H. F. MANSKE, F.R.S.C.

E. G. D. MURRAY

WE ARE NOT MASTERS of our own destiny because of our genetical inheritance and because our very existence depends on the interrelated activities of other animals, plants and bacteria. The accumulated knowledge of biology and other sciences allows us to modify inimical circumstances and to adapt other forms of life and their vital processes to meet our own uses and needs. This measure of control is incomplete because many essential contributors to our existence, or to our destruction, are independent of man, and because of the inadequacy of our knowledge.

Human intellectual developments impose greater complexities, and interferences designed for one purpose not infrequently have disastrous effects in other and sometimes more important essentials. Such effects are as evident in disease control, in management of natural resources, and in husbandry as they are in human political relations.

Scientists should be less timid; they should demand and be granted more influence in the use of their knowledge, more say in the determination of financial needs for scientific research, and some control over the implication of science for political expediency.

Time gives no quarter and so, in this exposure of the biological aspect of man's oecological dependence, to which he adds restrictive jealousy, partiality and greed, the exponents can touch on only salient features illustrated by a few revealing examples. Yet we may quote Robert of Brunne and say "wyys is that ware ys."

K. W. NEATBY

IT IS NECESSARY to introduce this subject with a few sweeping assertions that merit more consideration and, perhaps, qualification than is possible in one paper. The simplest living organism is an extremely complicated piece of matter, and the experimental biologist is confronted with so many interrelated variables that his conclusions with respect to behaviour under various circumstances are usually expressed in terms of probabilities. This is true if experiments are concerned with single species cultured under carefully controlled conditions. The problems are much more complicated when the purpose is to define the conditions that are favourable or unfavourable to a species in free competition with a host of others.

Of all organisms, man is the most difficult and, experimentally, intractable. He is the most complicated of all, the time between one generation and another is lengthy, and controlled experiments that imperil the life of an individual are seldom possible. Moreover, even if we could eliminate the varied and widespread objections to interference with man's reproduction, the problem of defining "fitness" might prove even more difficult. It is certain that we now know enough to alter the course of our own evolution should we choose to do so. Indeed, we are already doing so by preventing countless thousands of people from dying of diseases, some of which clearly involve hereditary predispositions.[1] Perhaps some of the many who stress the importance of philosophical and spiritual considerations relating to man's reproduction might well ask themselves if interference with natural death should not be restrained. Surely none can deny that it is natural to die of smallpox, diphtheria, typhoid, tuberculosis or any one of many other diseases. If we insist on the right to interfere with death and, at the same time, proclaim a hands-off policy with respect to procreation, we must be prepared to cope with the ecological problem we ourselves create. This, in my opinion, is the core of man's biological dependence.

Man is wholly dependent upon other forms of life for his existence, and if he is to survive, let alone continue to multiply at the currently prodigious rate, he must devise means to encourage species he regards as useful and discourage those that are directly or indirectly inimical to his own welfare. His record so far is a mixed bag. I hope to show that mismanagement of the biological elements of our environment is due more to our inability or unwillingness to apply the knowledge we possess than to ignorance, although application is not likely to increase much until

[1]Tage Kemp, "Medical Genetics in Recent Years," *Danish Medical Bulletin, III* (1956), 129–35.

knowledge is more widely distributed, and we still need to know much more. I know too little of agricultural history to pass judgment on the state of affairs in thickly populated countries where farming has been practised for centuries; but even allowing for a measure of extravagance in the claims made by conservation enthusiasts there are good grounds for careful consideration of the limitations of natural resources in North America.

All forms of life that have been studied, whether they inhabit communities undisturbed by man, farmers' fields, or orchards, fluctuate in abundance; some violently, some gently, some over long periods and some over short ones. In almost any community hundreds or thousands of different kinds live in a state of complicated interdependence. Just as changes in the environment usually confer a selective advantage on certain strains of a given species, so too they alter quantitative relations among different species of a community. Of course, deliberate or fortuitous introduction of new forms, or elimination of one or more, imposes new environmental factors on the inhabitants. I shall deal more fully with this subject in a few minutes.

Since the success, indeed survival, of a species, as of an individual, depends upon interactions between inherited potentialities and environments, digression at this point is inescapable. In cross-fertilized species, different individuals rarely respond identically to given environmental conditions. "One man's meat is another man's poison." Likewise, the impact of an environmental stress differs not only with different species but also with different strains and individuals. What I want to say about man's biological environment will be unintelligible to elements even of this enlightened audience without some reference to heritable variations.

Three major factors operate in evolutionary change. The first is mutation, the second is recombination, and the third is selective elimination. Mutation consists in alteration in a gene (besides more or less gross chromosomal changes) that results in a disturbance of some physiological process leading to a recognizable change in the individual carrying it. If a mutation seriously impairs the vitality of individuals carrying it, disappearance soon follows. If it does not, in a few generations it will be given an opportunity to express itself in a multitude of different genetic environments. In most forms of life the number approaches infinity. Cross-fertilization is an amazingly efficient means for providing an enormous number of different recombinations of genes. Natural selection is merely a matter of selective elimination and the competitive advantage conferred upon genotypes suited to the environment. It is a continuous experiment of an infinitely complex design. History never repeats itself exactly and nature is never perfectly balanced, at least not for long. As the defendant in "Trial by Jury" said:

Of Nature the laws I obey
For Nature is constantly changing.

In attempting to terminate my introductory ramblings and to come to grips with the subject assigned to me, I have wondered if it might not be equally or more appropriate to discuss the rewards of knowledge of man's biological independence. He is certainly the least dependent of all species on the globe in that he can deliberately alter the course of his own evolution as well as that of plants and animals on which he depends for survival, and he believes himself to be susceptible to a wider range of pleasurable sensations than any other organism. What then stands in his way? If our limited knowledge were more widespread, if it were supported by understanding and followed by courageous action, there are probably few obstacles that could not be overcome. We are pretty much in the position of the farmer who refused to buy a farmers' encyclopaedia on the grounds that his knowledge was already far in advance of his practice. However, our position can only be secured, if it can be at all, by persistent effort. Let us look first at food.

I have estimated that our consumption of food in Canada is currently equivalent, in terms of dollars, to not less than 80 per cent of our production. Production has increased during the past ten years, but this increase has been possible mainly through an unprecedented period of favourable weather. In comparison, increases due to the cultivation of additional land, improved farming methods, additional irrigation facilities and better crop varieties are small. The amount of good virgin farm land is limited and before long the new settler will be in direct competition with the forester. Moreover, the capital requirements for development are considerable. Improvement in methods of farming is, of course, possible as in anything else, though it will always be limited by the average capacity of farmers. Since the introduction of sprinkler irrigation, topography has become less significant as a limiting factor in available irrigable land, but water supply and power are not unlimited and equipment is expensive.

It would be hazardous to predict the outcome of modern methods of crop improvement. The introduction of hybrid corn, the direct result of genetical researches on hybrid vigour, has increased yields by about 20 per cent. This is probably the most spectacular success in the realm of modern plant breeding. The recent development of methods for transplanting to bread wheat intact chromosomes from related species and genera has opened a new chapter in wheat breeding. This, combined with increasing the natural rate of mutation by use of ionizing radiations, enormously widens opportunities for useful selection. However, it would be rash to expect increases in wheat yields approaching those secured in corn, and any increase that does not entail some sacrifice in quality may be extremely difficult to get.

Plant-feeding insects and plant diseases are powerful competitors with man for food. In Canada, with one or two doubtful exceptions, we have yet to eradicate a single one. Every crop plant is subject to damage by several to many insects and a variety of diseases caused by fungi, bacteria, viruses and other agents. Control is effected mainly by development of resistant crop varieties, by use of toxic chemicals, by biological control agencies such as parasitic or predacious insects, or cultural practices, or by some combination of these methods.

We have learned from experience that disease-resistant varieties of crop plants usually enjoy only temporary security. The pathogen can too often say to the host what the sergeant said to the pirates:

> To gain a brief advantage you've contrived,
> but your proud triumph will not be long-lived.

This arises from the fact that the capacity for genetic variation in pathogens may be little less than that of hosts. Experience has shown only too clearly that introduction of a new variety of a crop plant resistant to a particular fungus disease is likely to be followed sooner or later by appearance of a race of the fungus capable of parasitizing it.

The entomological picture is similarly complicated. Just as bacterial pathogens of humans can evolve strains resistant to antibiotics, so have a notable number of insects responded to exposure to toxic chemicals.

We alter the course of evolution of parasites by changing their environments. We have achieved important successes, especially with fungus diseases, though most have proved to be temporary. Progress in the control of virus diseases is less encouraging, partly because genetic variability in host reactions, on which breeding for resistance depends, is much less common. Notable progress has been made in the development of insect-resistant crop plants, but economically important examples are few and there is good reason to expect that plant-feeding insects are, in general, able to adapt themselves to uncongenial hosts as readily as fungi. In this realm of crop protection, the security we now enjoy will probably not be maintained, let alone strengthened, without an increased understanding of the genetics and physiology of hosts and of parasites, and of host-parasite relations. At present our knowledge, though important, is fragmentary and elementary.

From the point of view of plant protection, our debt to the chemist is immense. Few crops are grown, at least in North America, without the protection of synthetic chemicals. But these controls are troublesome and expensive, and too often entail undesirable secondary effects. The chemist moves faster than the biologist. His aim is to produce something that will protect economic plants and animals from damage by a particular organism or orgnisms without injuring the plant or animal to be protected. Some modern fungicides and insecticides are highly selective and we have reason to hope that as we learn more about the way the chemicals produce their lethal effects still higher degrees of selectivity may be possible. Nevertheless, the use of toxic agricultural chemicals often leads to unexpected and, usually, unwanted effects. For example, the problem of the oystershell scale on apples in Nova Scotia was virtually created by use of sprays containing mild sulphurs for controlling the fungus disease, apple scab. Prior to 1930 the scale insect was not a serious problem, but later it was responsible for very serious economic damage. Investigations revealed that this insect could be kept under control if its two most serious enemies, an insect that parasitizes it and a mite that preys upon it, were not interfered with. Mild sulphurs were toxic to them, but when fungicides harmless to the mite and parasite were used for controlling apple scab, scale populations decreased and this insect ceased to be a serious economic pest. A similar example is provided by the well-known increase in abundance of the European red mite following use of DDT for controlling various orchard insects. This sort of thing is not surprising. An apple tree normally harbours at least a hundred species of insects and other arthropods. Some are plant feeders, some feed on the plant feeders and all are subject to parasitism, predation or both. It should be clear, therefore, that the intelligent use of pesticides depends upon knowledge of extremely complex communities of living things.

Changes in cultural practices are also likely to be followed by new problems or aggravation of old ones. Strip farming in Western Canada was adopted as a measure to control soil drifting. It provided almost ideal conditions for the wheat stem sawfly, which overwinters in stubble. The cropping system is mainly alternate wheat and fallow. With strip cropping, none of the wheat is more than a few rods from last year's stubble. Fortunately the development of sawfly-resistant wheat varieties has provided a partial answer to a problem in prairie wheat production exceeded only by drought and rust. In the interior of British Columbia, scab and powdery mildew of apples have both increased in importance recently. There is reason to believe, though opinion is divided, that sprinkler irrigation provides more suitable conditions for these diseases than does flood irrigation.

I hope I have said enough to show that the problem of plant protection is a complicated one. Nearly all our cultivated crops, other than trees, have been introduced from elsewhere. Many of the most important plant pathogens and destructive plant-feeding insects and most of the important weed species are immigrants. Each year brings with it new crop varieties, new toxic chemicals and new methods of management.

No doubt, despite all hazards, we shall continue to increase food production in Canada for a long time to come. It is unlikely, however, that the increases will keep pace with the population. The pinch will probably first be felt not in how much there is to eat, but in what there is. A given acre of land will produce about five times as much nutrient energy in the form of plant products as it will of animal products. If we are prepared to give up beef, bacon and butter and content ourselves with a diet of cereals, soybeans and algae, highways and beaches can become much more crowded before we starve. This, I have no doubt, is what the manufacturers of automobiles, television sets and deep freezes can look forward to with pleasure.

No matter how successful we may be in food production, sooner or later we shall find that good land is limited in quantity. We all know what is now happening around Lake Ontario. The same thing can happen on the Saanich Peninsula, in the lower Fraser Valley, at Edmonton, Winnipeg and elsewhere. Cities and industry are developing on the best agricultural land. I do not know how much farm land in Canada has been lost in this way and I doubt if anyone else does. Even though the amount is small in relation to the total, as it certainly is, it might be well to examine the possibility of encouraging future urban, suburban and industrial developments on land unsuited to agriculture.

Human society appears to be regulating its affairs on the assumption that abundant food, a minimum death rate and peaceful settlement of international disputes will guarantee our physical well-being indefinitely. Our actions will probably continue to be governed by this fallacious assumption for a long time to come; and since the biologist has, perhaps, rather less influence on international affairs than the plumber and certainly less than the lawyer, he can probably serve best by devoting himself to food and other aspects of human health and, at the same time, indulge his fascination in life in its many and varied forms.

It is fitting to discuss briefly two additional subjects that are becoming increasingly important in relation to health and about which too little is known. First I shall discuss briefly the problem of toxic residues, an inevitable aftermath of the chemical control of agricultural pests, and then I propose to touch upon the subject of radiation hazards.

In North America very few crops can be grown successfully without the protection of pesticides. Terms such as fungicides, bactericides, insecticides, acaricides, nematicides, and herbicides are commonplaces among farmers. Various chemicals are also used widely for the protection of stored foods, for the protection of farm animals against many different parasites, and for controlling biting flies and several arthropod vectors of human diseases.

Some of the agricultural chemicals in use today, such as the organic mercurial seed dressings and parathion, have been responsible for severe illness and even death; but all cases have arisen from carelessness. Many other materials that are not likely to give rise to acute illness or fatal poisoning are detectable in a variety of food products. As yet there is no evidence that anyone has suffered illness due to the ingestion of the small amounts of poisonous residues found in food as a result of the normal agricultural use of pesticides. It is very difficult to define the magnitude of this hazard, because data on toxicity derived from experimental animals may not be applicable to man and assay methods, both chemical and biological, leave much to be desired. Moreover, we lack adequate clinical methods for detecting distinct syndromes associated with toxic effects of pesticides. It has been shown that DDT can be accumulated and stored in animal fat and that it and other chlorinated hydrocarbons may be eliminated in milk fat. Since some of these substances are more toxic to young mammals than to adults, their presence in milk should not be tolerated. Reconciliation between the needs of agriculture and the caution of medicine can only be achieved by improved knowledge and by techniques more refined than those now in use.

There is still a great deal to learn about the hazards of ionizing radiations, both qualitatively and quantitatively. The most serious long-term aspect of the problem is that of mutagenic effects. Muller[2] has calculated, on the basis of natural mutation rates, that at least one person in every five, on the average, carries a mutant gene not carried by his or her parents. It is clear that unless elimination balances occurrence there will be an accumulation of mutant genes, most of which are damaging. Avoidance of exposure to mutagenic rays is, therefore, important particularly for persons who have not passed the reproductive age. Fortunately, experimental atomic and thermonuclear explosions do not materially increase our exposure to gamma radiation. Increasing awareness, on the part of radiologists, of the genetic dangers of exposure to X-rays will probably result in greater protection from this source of danger for our children than we have enjoyed.

Increased mutation rate is by no means the only risk attendant on

[2] H. J. Muller, "Genetic Principles in Human Populations," *Science Monthly,* LXXXIII (1956), 277–86.

exposure to ionizing radiation. Warren[3] has made a study of some 82,000 physicians who died during the period 1930 to 1954. The average age of death of the radiologists was 60.5, of the others 65.7. This difference was not associated with any specific effect, but death from all causes occurred, on the average, at an earlier age among the radiology group.

If atomic explosions are not carried beyond the experimental stage, the greatest danger appears to be associated with the fall-out of strontium-90, a radioactive isotope of strontium with a half-life of about 28 years. In absorption by plants from the soil and in animal metabolism, this element tends to follow calcium. Despite preferential selection of calcium by plants and a measure of selective elimination of strontium by animals, minute amounts of strontium-90 are being steadily deposited in the bones of all of us, particularly in the young. At a certain level, strontium-90 promotes the development of bone cancer. It does not occur naturally, but on the average we now carry in our skeletons about 0.12 micro-microcuries of it per gram of calcium.[4] Some of us contain ten times this amount. It reaches us mainly through vegetables and dairy products. The average figure I have quoted is considered to be only about 1/10,000 of the maximum permissible concentration, but without further experimental explosions we can expect to harbour about ten times as much in another ten or fifteen years.

X-rays and atomic explosions are not the only radiation risks confronting us. Consideration must be and, indeed, is being given to hazards attending the development of atomic power and the disposal of atomic wastes. The penalties of ignorance in this branch of biology may be serious.

There are many other aspects of man himself and his environment that might serve to illustrate the complexity of our biological dependence quite as well as those I have employed. It is true that by virtue of special gifts and experience we have achieved an amazing state of security and comfort in, on an evolutionary time scale, a very short time. If what we are pleased to regard as progress in population and industry continues for another fifty years or so, will it be automatically followed by equal or better security and comfort? Judging from concern for biological affairs in North America, we take this happy future for granted. According to a recent study, "Less than about 10 per cent of our total research

[3]Shields Warren, "Longevity and Causes of Death from Radiation in Physicians," *Journal of the American Medical Association*, CLXII (1956), 464–8.
[4]J. L. Kulp, W. R. Ecklemann, and A. R. Schubert, "Strontium-90," *Science*, CXXV (1957), 219–25.

expenditure goes into biology and medicine."[5] This statement relates to the United States only. In Canada it is probable that a larger percentage of our research effort is devoted to biology if only because industrial research is much less prominent. However, our total biological effort may look, in relation to the problems discussed here, much better than it is. It is mainly devoted to applied work in medicine, agriculture, fisheries and forestry. A limited amount of creditable fundamental research is associated with applied work, usually being justified by the needs of the latter. This, in my opinion, is clear evidence of a deplorably unenlightened attitude on our part. It is bad enough that science students should "take" English only to enable them to meet the dubious standards of scientific journals, but perhaps worse that they should assume a similar attitude towards biology only to qualify as applied scientists.

In order to close on a brighter note, though I dare not regard it as prophetic, I quote from a recent paper by Beadle.[6]

Man's evolutionary future, biologically and culturally, is unlimited. But far more important, it lies within his own power to determine its direction. This is a challenge and an opportunity never before presented to any species on earth.

It has been clear for a long time that man is potentially capable of cultural self-direction—that he could, to a much greater extent than he now does, consciously select his cultural objectives. What is not so obvious is that it has now become possible to exercise a comparable degree of control over his purely biological evolution.

Through the understanding of heredity that man has gained within the past half-century, he has acquired the power to direct the evolutionary futures of the animals he domesticates and the plants he cultivates. At the same time and in the same way, he has won the knowledge that makes it possible deliberately to determine the course of his own biological evolution. He is in a position to transcend the limitations of the natural selection that have for so long set his course.

IAN McT. COWAN

MAN, LIKE ALL ANIMALS, is a member of the highly complex ecosystem in which he lives. Many of the living components of such a system have the capacity of modifying their environment, of making it, by their very presence, more suitable or less suitable to support others of their kind. Man is outstanding in the category and is unique in his ability to choose the nature of some of the alterations that his presence

[5]"Preliminary Report of A.A.A.S. Committee on the Social Aspects of Science," *Science Monthly*, LXXXIV (1957), 146–51.
[6]G. W. Beadle, "Uniqueness of Man," *Science*, CXXV (1957), 9–11.

engenders. There are, however, very large areas in ecosystem dynamics in which man exerts no mastery of direction, velocity or ultimate effect of the changes wrought by his presence. The direction taken by the forces of change are frequently positive at low densities and negative at high densities, and the point of inflection is most difficult to determine.

In Canada today we ignore the existence of such an inflection point. At every turn we hear men of commerce extol the virtues of greater population. Our national policy is directed toward increasing the number of men in Canada as rapidly as possible without much regard to the biological forces involved. Even if we desired, it is doubtful that we could control the increase in our population. It is part of a world trend and the forces are at the moment beyond human influence. I am not convinced, however, that the increase in numbers will continue more or less indefinitely as has been assumed in so many studies of human population. Predictions in human demography have been notoriously uncertain. The present rate of human increase is quite exceptional but even it has been far from uniform. In some areas population has been stable or decreasing, while in others the rate of increase has been very high.

Goodhart (1956) has shown that there is selective advantage in high fecundity in primitive cultures, but that this advantage is lost under certain improved social conditions, and under conditions where food rather than disease is the factor limiting human numbers. As these circumstances arise in more and more areas of the world, we may—and probably shall—experience important changes in trend.

Be this as it may, we must plan for much larger populations before we learn to control our enthusiasm for unbridled procreation. Dr. Neatby has dwelt briefly upon some of the biological problems presented by agriculture under such circumstances. My concern is with the native animal resources, the part they must play in our future and the fundamental understanding we have gained and have yet to gain about them.

In any country the primitive position of the wild animals (including fishes in this category) is that of a basic resource for food, clothing, tools and transport. The advance of settlement through the pioneering phase to stable agriculture, urbanization and industrialization brings accompanying changes in the concept of the use of native fauna. Certain species become recognized as of continuing primary value for food or economic gain, others become largely of sporting interest, still others may escape their natural controls and become important impediments to human activity. Man finds also that there is much that he can learn from animals that will contribute to his understanding of himself and of his adaptations to environment. Flight, camouflage, hibernation, adaptation to cold, factors of growth, hormone function, and many other aspects of the life we share, can best be understood by the study of animal forms. Finally man comes to take an interest in much of the fauna for itself. He finds in living animals a source of interest that vitalizes his enjoyment of his environment. In Canada all stages of evolution are to be seen contemporaneously. In the far north, there are still populations living directly off the native animal resources. Further south, native stocks are used as temporary and local contributors to human subsistance, while on the edge of urban areas, except for commercial fisheries, the use is wholly recreational.

Few realize how closely interwoven are the lives of men on this continent and the native wild vertebrates. Lampreys, carp, pike, dogfish, hair seals, sea lions and several other species reduce man's catches of desirable fishes or increase the cost of operations. Deer, field mice and insect pests invade his orchards; mountain lions, wolves, coyotes, foxes and lesser carnivores raid his herds and flocks; bears destroy his apiaries, his livestock and sometimes his forest plantations. Mallards and pintail ducks raid his grain fields, finches eat the buds off his fruit trees, while robins eat the fruit. Insects, porcupines, squirrels, mice, rabbits and deer harass the silviculturist. Furthermore, wild animals and birds can serve as sources of parasites and disease transmissible to man and his domestic animals.

On the other side of the ledger, we in Canada annually take as many as seven million fur skins from twenty-four species of native land mammals; we kill, and use for their meat, bones and oil, between four and five thousand whales. The annual harvest of seals exceeds three hundred thousand individuals of six or seven species. Shrews, red-backed mice and white-footed mice play an important role in curbing outbreaks of certain forest insects. From thirty to sixty thousand caribou annually help to feed and clothe northern human populations, while more than a quarter of a million large game mammals of ten species are used as food throughout the nation. To this must be added the two million waterfowl and one and one-half million upland game birds annually used by Canadian sportsmen. These two groups alone comprise forty-two species. At least nine species of purely marine fish support an industry annually yielding eight hundred thousand tons of food to the world's burgeoning population. In addition, eight species of anadromous fish make a further contribution of ninety-six thousand tons.

All these animals, the biologist must understand in sufficient detail to permit man to discourage those inimical to his interests, to encourage those he desires to use. The task is a stupendous one, but it is by no means the most difficult one before the biologist, for, as any scientist working in the applied field will confirm, an equally demanding part of his task often comes when the research is completed and he must translate the critical parts of his findings into terms understandable by the biologically uninformed and persuade them to adopt an appropriate management policy.

Apart entirely from the commercial and semi-commercial interest we have in our fishes, birds and mammals, we must not neglect the host of species that are as much a part of Canada as her lakes, mountains and prairies. These are an essential part of our environment and it is the duty of the biologist to maintain them, to see that none pass from the earth through the instrumentality of those whose only approach is wholly contained in that question, "what good is it?" As has been so well said by Maurer, "The majority of great writers, many abstract thinkers, the greatest scientists have avowed an intimate need of nature. In some of them, the thirst for natural things, for the full sky, landscapes, trees, flowers, wild animals, the tang of the autumn wind, the tumbling seas and tranquil lakes has been an obsession. They have truly fed upon nature in all its aspects. The implication is clear that severed from nature man's imagination and inquiring mind would diminish, perhaps wither utterly."

Some pattern for the future can be seen in the comparison of the use of native animals by human populations of widely different density. This use is conditioned as much by cultural background as by density, and we must therefore seek comparison in areas of similar background and approximately similar ecology.

Present human densities in Canada vary from forty-five per square mile in Prince Edward Island and thirty-one per square mile in Ontario, through the western provinces, where there are between three and four people per square mile, to the Northwest Territories with .015. The mean density over Canada is about four per square mile. Under these circumstances, terrestrial wildlife is producing about forty-eight million pounds of edible meat per annum and something under seven million salable fur skins. The production of wild animal units per capita is about 0.3. Correspondingly, Finland, with a density of thirty per square mile, produces 0.3 animal units per capita, and Denmark, with a density of two hundred per square mile, still yields 0.6 wild animal units per capita. This is about double the productivity of Canada today. The species of course are very different and comparison must be made with caution.

These figures are quoted to show that in northern communities of our social background and ecology, we need not expect to write off the use, and much less the survival, of native terrestrial game mammals and birds as human population densities reach figures many times greater than those now present in Canada. However, it follows clearly that as greater demands are made upon our usable wildlife resources, we will have to learn to manage the stock to much finer tolerances. This we could not do armed only with the information we now possess and within the administrative framework currently established. Indeed our knowledge is already inadequate for the task in hand.

Across the vast, rugged northern two-thirds of Canada, wildlife, as a crop, provides the highest yield per square mile that such land can produce under present and foreseeable economic conditions and under the foreseeable needs of man. In the animals and plants of this area is the key to human development of the Arctic. The productivity of this land can be maintained, but to do so will require more intensive and imaginative studies of the ecology of the animals concerned than any conducted so far. The first essential for the development or maintenance of a native animal resource is the creation or preservation of habitat lying within the range of tolerance of the species. It cannot be truly said that we know these limits accurately for any wild species. Beyond this we deal with animal populations. Already much is known of the actualities of population instability and of the most important forces contributing to changes in population size. We know much of the interaction of prey and predator, of density and disease. We have much suggestive evidence that stress, of a psychological nature, has been woven into the fabrics of many species of birds and mammals, as a factor limiting ultimate density.

In Canada, more money and effort has gone into the study of pheasants, waterfowl and barren ground caribou than of any other terrestrial species. We have sought to encourage their reproduction, to supplement their populations, to protect them from their predators, but with all we have found out we must still admit that the productivity of these species largely defies our control, and we are still almost powerless to protect them from their enemies. Yet they are, and will continue to be, species of the greatest importance to Canadians. We have made a good start in that we now realize that there are no simple solutions; answers must come from fundamental studies which cross the boundaries of all scientific disciplines.

The barren ground caribou is an excellent example. This large deer is the mainstay of survival for the natives of Arctic Canada and is of the greatest use also to the resident white man of the hinterland. An extensive survey of the population in 1950 revealed upwards of half a million animals, maybe as high as six hundred thousand animals. Five years later a survey showed that the stock had fallen by one-half. The essential elements in the decrease are reasonably apparent: an effective birth rate of between 7 and 14 per cent overwhelmed by an annual mortality of almost double that. The mortality of adult caribou is apparently contributed to in almost equal amounts by timber wolves and by man, with an unknown proportion taken by other natural forces. Thus the immediate palliatives are apparent: fire protection, control of predatory animals and restriction of human use to the minimum. The big question surrounds the low birth rate, which should not be less than 30 per cent. The exploration of this aspect is in progress.

The matter of wildlife damage to agricultural crops is already vexatious in large areas and will inevitably become more so as exploitation of still more marginal land brings agriculture closer against the wilderness haunts of wildlife. Main damage will be by deer and waterfowl. Here again we are limited by ignorance. Short of killing the offending creatures, we do not know how to discourage them from attacking our crops. Unless effective techniques can be developed, as the pressure for human food increases, it is inevitable that the native animal stocks that compete with man for our agricultural crops will be reduced. It is incumbent upon those groups in our population primarily interested in these species as a recreational resource to support more active research into the problems presented and to press for the translation of facts already known into effective management policies.

What I have already said of our knowledge of the terrestrial animal forms is almost equally true of the aquatic ones. Here, however, there has possibly been less lag between discovery and application. We know quite a lot about the general behaviour of fish populations, but next to nothing about the individual fish, how it is adapted to its environment, what the stimuli are that activate its various sense organs and how it responds to these; what its thresholds of tolerance are for the various natural forces that limit its survival, distribution growth and reproduction. There is little to be gained by dwelling on past mistakes unless we can gain profit from them. With this in mind, it is pertinent to point to the catastrophic results to the Great Lake fisheries that have arisen as a result of introducing the sea lamprey into the Lakes. The introduction was an inadvertent consequence of the construction of locks and canals that eliminated the natural barriers. There is real danger now of a similar situation arising in some of our finest Pacific coast salmon-producing rivers through the diversion into them of waters from the pike-infested rivers of the Arctic drainage. In the light of present knowledge, this would be inexcusable folly.

The urgency of our need for a more widespread understanding of long-term consequences of human activity is probably more clearly apparent with respect to the anadromous Pacific salmon than to any other fish. The great and small rivers of the Pacific drainage of North America are the nursery areas of vast stocks of salmon. They also offer most attractive opportunities for developing hydro-electric power to satisfy the increasing needs of a power-hungry civilization. The question immediately arises: is it necessary to sacrifice one of these benefits to humanity in order to develop the other?

Even though there is a world shortage of food, in Canada today there is a surplus, and it is tempting to some to neglect the long-term primary importance of a food resource in their approach to the decision of priority use of a river. To the biologist, however, it is the height of folly to permanently destroy any source of food, unless a greater source is thereby developed that could not have been brought into production in any other way. The second important question that arises in considering alternative uses of a river is: if it is not a direct alternative, to what extent can both uses be maintained coincidently? Successful, coincident use of a resource is an ideal well worth striving for.

The Fraser River is the largest producer of salmon in North America, if not in the world, and can be used as an example of the biological problem arising. The development of the full power potential of a river with fluctuating flow is dependent upon the construction of one or more dams behind which water can be stored during peak flow, to be released during slack periods. The dams constitute barriers to the movement of fish up and down the river. Furthermore, the impounded water areas profoundly alter the delicately adjusted ecological characteristics of the river system. This alteration may be as lethal as complete destruction.

The problem of fish protection centres on the free passage of the fish as the adults move up river to spawn and again as the smolts migrate down river to continue their growth in the sea. The fish traffic in the Fraser is prodigious. In a year of maximum production, one large tributary system of the Fraser alone sent down the river two hundred million migrants of but one of the three species using it. The upstream movement in the same system involves the passage of as many as two million fish during a single season. Eight million are caught commercially. Four of the five species of Pacific salmon ascend the river far enough to require consideration when making decisions on power development; however, only the sockeye and spring are of major importance.

Too many biological integers must be inserted into the equation for consideration of all of them here, but some of the most important ones can be reviewed as examples of the type of biological problems that demand answer.

The motivation of migration of both smolts and adults and the stimuli that direct their passage are among the most critical points to be considered. A clear understanding of the stimuli, chemical or physical, that permit the migrating fish to direct its passage is an essential precursor to full development of the potential of any salmon-producing river. Many types of fish ladders and of fishways have been constructed, fish elevators, tank trucks and other ingenious devices have been designed, but basically their operation depends upon the attraction of fish into them. The selection of the home stream as the final goal, and the ultimate choice that leads so many of the adult migrants to spawn in the stream in which they themselves were hatched, must almost certainly involve an olfactory response to which the fish was sensitized in the nursery area. What then is the nature of the stimulus? Is it possible to use the stimulant, or to substitute for it some man-made chemical that can be used to lead fish into our passage devices or up unused streams to man-made spawning beds?

Another factor in the upstream journey is the ability of the salmon to perform. It does not eat after it reaches the mouth of the river and it must move up the river—sometimes for some days and possibly hundreds of miles—with an energy budget as clearly limited as a single tank of gasoline. Like an athlete, it can jump only a certain height, swim only a certain speed and negotiate only so many jumps and runs without tiring. The designs of fish ways must take cognizance of these physical limitations. Where a dam is so high that the fish cannot surmount the lengthy series of runs and leaps imposed by a fishway, locks or lifts can be substituted, but the frantic struggles of the fish to escape from these devices often result in the same endpoint of exhaustion. Always, it must be remembered that the energy reserves are limited and are not being replaced. The fish starts up river with reserves designed through evolutionary experience to bring it to the spawning beds with energy sufficient to accomplish the act of reproduction. The energy account cannot be overdrawn; delays or other environmental changes that squander the energy are as fatal as the fish net.

The downstream movement is apparently directed by visual and current clues and presents problems of even greater magnitude than the spawning migration. These are the kinds of problems we must answer. We cannot neglect or ignore them and proceed to destroy resources that to us now mean millions of dollars in annual revenue, but to those Canadians yet unborn will mean high protein food in a world in which such is desperately scarce.

It is not by physical obstructions alone that we can interfere with productive capacity of our river systems. Recent studies have shown that salmon are so sensitive to certain organic chemicals that may be introduced into the water that dilutions of as little as one part in one billion are sufficient to act as violent repellents. It is imperative that man learn how to dispose of his waste products without disturbing the critical biological adjustments of other important animals of the environment. As Hoar has so well stated:

Like all animals, salmon can be readily exterminated if their environment is rapidly altered beyond certain rather narrow limits. On the other hand, salmon like all other animals have a certain plasticity, an ability to adapt and to evolve at a slow rate within certain definite limits. If it were, at the moment, possible to state for each kind of fish and for each stage in its development these critical limits, for adaptation and for evolution, then it would be simple to predict the salmon's fate in a changing environment. Indeed for each and every contemplated change in the river systems it would be possible to calculate the biological changes just as it is possible to calculate the physical changes.

At the moment there are numerous assumptions concerning the biological effects of water utilization but few facts of the kind required to make statistically significant predictions. If man wishes to maintain the populations of fish and also to utilize the waters in which they swim, he must establish the limits for healthy survival, for reproduction, for adaptation and for evolution. In contemplating this, he must recognize the complexity of biological material and be prepared to spend time, effort, money and human ingenuity on problems which may prove far more difficult than the power development itself.

Public apathy and ignorance already prevent the proper application of what we know. A wider public understanding of the biological forces at work in man's environment is urgent, along with an appreciation of the complexity of living material and a willingness to use biological advice more freely in guiding human activities that influence the renewable resources. At the same time, our biological understanding is far from perfect and demands greatly increased research effort. The easy and spectacular things have been done, it is time now to concentrate on the deep-searching, fundamental studies that will give us the understanding we must have if we are to reap the full benefit that the native animal resources have to offer.

Our fishes, birds and mammals in their native habitat have as much to give to our future as have any of our other resources; they are a perpetual source of the energy that is life, but they are dependent upon our wisdom and foresight for their survival. We must do all in our power to perfect our understanding of our environment, to prevent the well-meaning but misguided from teaming with the rapacious and self-centred to the destruction of large parts of our animal resources. We must translate to the people and to the governments the high purpose of resource use with an abundant future as the goal.

G. H. ETTINGER

THE BIOLOGICAL PURPOSE of life is to survive. Man's survival depends on his ability to adapt to his environment, normally considered to be the external conditions in which he lives. But physiologists accept the view of Claude Bernard, that we have an internal environment and that, in his words, "All vital mechanisms, however varied they may be have but one object—the preservation of conditions of life in the internal milieu." The internal environment is that space between the integument and the lining of the alimentary tract, in which the organs which maintain our life discharge their functions. Here tissue fluids carry oxygen, simple foods and waste substances; transport hormones and protective antibodies; lubricate joints and muscles; favour the mysterious movements of nervous impulses, the creation of thought and the filing of memories. Temperature, fluid volume, mineral, hormonal and food content must be kept within normal limits in order that life may proceed smoothly; the constancy of the internal environment—or in Cannon's term "homeostasis"—determines health. Normal growth depends upon it, and the proportion of hormonal content at particular phases determines attitudes, actions, habits, and emotions. "Physically mentally, sexually, emotionally, we are largely the product of our hormones," says E. V. McCollum. Exhaustion or excess of a constituent may create an urge or appetite, leading to positive action in the external environment—thirst, leading to drinking; air hunger, leading to excited respiration; chilliness, leading to muscular activity or added clothing sexual desire, leading to overt activity.

There is no part of man's biological dependence of which there is greater ignorance than of his internal environment. We can measure his blood-sugar and determine that he needs insulin. We can X-ray his chest and learn that he has tuberculosis. We can examine his hair and find that he has been poisoned by arsenic. And we can, with this knowledge, mend him, for we can understand the processes which underline these illnesses. But the sum of mysterious forces that lead to behaviour are still recognized only in their effect. The adolescent child becomes a problem to his parents; he is under the influence of maturing hormones and myelinization of nerve fibres. He reaches college age and finds that

> Not a girl goes walking
> Along the Cotswold lanes
> But knows men's eyes in April
> Are quicker than their brains.[1]

Too often this mysterious force leads him to a marriage for which he is too young, to one who is incompatible, and ends in impoverishment and disaster. Is this purely an hormonal effect? We do not know. Castration after marriage does not invariably destroy sexual activity.

But much more tragic is our ignorance of the processes of the mind, and our difficulty to forestall or treat, except empirically, mental illness. So widespread is this affliction that the care of the mentally ill is regarded as a social responsibility, and governments build huge hospitals staffed with physicians especially trained to give total care to these unfortunate people. The penalty of man's ignorance of his internal milieu results in our inability to cope with the greatest and most universal scourge—the sickness of the mind.

Fundamentally, man's survival depends on his ability to adapt his life to changes in his external environment. Indeed, Herbert Spencer defines life as "A continuous adjustment of internal relations to external relations." This adaptation includes the ability to acquire food and protection and to defeat his enemies. Man rises above the lower animals in his powers of creation. Placed in an inhospitable environment he devises simple instruments to help him obtain food, clothing and shelter. He grows up in a community of his own folk, mates (with or without the ceremony known to us as marriage), and raises children. His group exhibit characteristics of bodily contour, pigmentation, language, habits, dress, and family life by which they are recognized as a race. The racial customs are sufficient for survival, provided the climatic conditions are unchanged, the cycle of animal life about him is undisturbed, disease is not introduced, and his privacy is not invaded by racial strangers who cause him to deviate from practices which for centuries have maintained a healthy perpetuation of his race. War with his neighbours, a rigorous climate and periods of famine keep his numbers fairly constant. If food

[1] John Drinkwater, "Cotswold Lanes."

is chronically scare, infanticide is an understandable custom, and the destruction of the elders who can no longer contribute to the larder, excusable.

The survival of a race which has lived under primitive conditions, when it is invaded by a people who have reached a higher civilization, depends on the ease of its adaptation. Thus, in the Canadian Northwest, the Indian lives in squalor, "shiftless" according to our judgment, content with poor shelter and "unbalanced" food, his inclination to hunt frustrated by the poor hunting that the white man's progress has precipitated. For brief employment he can obtain supplies which last him for many weeks; he loafs and is weakened by the diseases which the white has introduced, but to which he has little resistance. His adaptation is poor—he will disappear. On the other hand, the Eskimo, intelligent, industrious, friendly, exhibiting a natural genius in handling mechanical devices, tends to adapt more easily. His great disability is the ease of infection with tuberculosis; at least one sanatorium in Ontario is kept open only because of the stream of Eskimos who come to it. If this weakness is controlled, the Eskimo will survive, though intermarriage with whites will soon make it impossible to find one of pure blood.

The higher the intellectual development of the race, the more complex are man's creations. The manufacture of and use of his inventions leads to the development of large communities whose residents depend increasingly on the conveniences provided. They have become adapted to a life in which their resourcefulness is limited to the narrow field in which they make their living. Confronted with a suspension of electrical power, a flooding of their land, a strike of a large labour union, they are reduced to ineffective improvisation, interruption of productive labour, exhaustion of food supplies, risk of grave illness, and a pathetic call upon Government to provide succour. They are biologically dependent on the mechanical support of their own devices.

But it is not only this dependence which places a narrow margin between effective life and survival. The process of manufacture of these devices introduces hazards to life. The miner may develop silicosis or pulmonary cancer, the factory worker poisoning with lead or organic solvents, the paper maker deafness. Exhaust from factory and mechanical vehicles probably contributes to the development of bronchogenic carcinoma. Noxious wastes are dumped into lakes and rivers; pollution, increased by sewage, is a hazard to him who drinks or bathes in the water. The careless disposal of waste products which has characterized industry leads one to suspect that the generation of power from atomic energy will create a potential hazard to those who live within a few miles of the installation.

With increasing specialization in industry and in business and professional life, there is evolved an increasing demand for social security, and nationalization of medical and hospital service. The threat of illness or unemployment to a population who depend on peculiar skills for their ability to purchase the necessities of life leads to demands for protection which is possible only by high taxation. The willingness to delegate responsibility for comfort and employment to Government and trade unions softens a nation.

This softening is personal as well as national; it is reflected in a falling off of physical fitness as well as of social independence. We take even our sports casually, preferring watching to participation. A Canadian professional hockey player who had played in Europe against the French, the Swiss, and the Russians praised their physical fitness, and said of the Russians that the most astonishing part of their performance was the frequency with which he met the same player when he was rushing from one end of the rink to the other. The American champion at billiards, Willie Hoppe, when asked why he did not drive instead of walking from his favourite billiard parlour to Madison Square Gardens to see the fights, said "It softens the legs." But the school boy does not walk, the college youth drives his own motor car, and the young man summoned to serve his country is physically undeveloped. We lack a programme of vigorous and compulsory physical exercise in Canadians of school and college age. It may be true that successful performance of work demands unusual fitness and skill of only that part of the body that is used in the work, and that ability to endure hard physical labour is no prerequisite for satisfactory production. But we must remember that a people addicted to softness of living may, in these days, be suddenly immersed in the hardships associated with mass destruction, and compelled to take the offensive against troops less accustomed to luxury and leisure.

An army about to attack the enemy prepares the medical corps for casualties, and estimates the cost in life. A fraction of its strength is said to be expendable. Major advances in invention, and in the use of new instruments, carry a similar expectation. It is the price we pay for their perfection. A test pilot is paid a large salary to compensate for his hazard. Accidents in air, water or land traffic kill people. Investigation reveals weaknesses in construction of the machine or judgment of the pilot; changes in design or regulations follow. Every year 3,000 people are killed in Canada in motor accidents; this is due less to defects in the design of motor cars, than to weakness in the driver or unwariness of the pedestrian. While a complacent public permits all who can obtain a licence to drive a car under current traffic regulations, on roads so congested in town and highway, we must consider these 3,000 people expendable.

We are all excited about the possibility that the testing of atomic weapons and the setting up of installations using nuclear reactors will pollute the air and reduce expectation of life, or induce genetic disturbances. Leukemia has occurred among the survivors at Hiroshima and Nagasaki, as well as among radiologists;[2] we have learned that life may be destroyed by man's experiments into the nature of things. Yet life itself may have resulted from the effect on simple organic compounds of these lethal rays, as has been suggested by the experiments of Paschke, Chang and Young[3] who synthesized amino acids by the action of gamma radiation from Cobalt 60 on ammonium carbonate.

But man *will* harness the atom, and he will make much greater use of it in the release of power than in the treatment of illness and the production of weapons. The lives that are destroyed or damaged in these years of experimentation are expendable—they are part of the price of future safe and effective use. Every discovery in medicine, tested first on animals, must produce experimental casualties in man. Sir Victor Horsley, chided for having killed the first few patients in operations for removal of the pituitary gland, said that these people died that other people might live, when the technique was perfected. The sulphonamides, introduced as bacteriostatics, saved many lives before it was discovered that in some patients they caused a fatal agranulocytosis. Penicillin, which succeeded the sulphonamides, was found to induce a sensitivity in certain patients, producing very serious results when the dose was repeated. Years of use of the natural antibiotics have shown that an organism previously susceptible may develop a tolerance, and, becoming resistant, may multiply in hospitals with a virulence hitherto unknown. A sad result has occurred in Canada—the danger of staphylococcic infection in our hospitals is greater than before the introduction of penicillin. Nature resents man's interference with her own control of life, and we pay the penalty of disputing her practices in the name of medical science. The introduction of Salk vaccine in Canada has had no tragic overtones such as occurred in the United States with the use of a preparation containing live virus, but many virologists doubt the wisdom of vaccinating with killed virulent organisms, lest there be a considerable survival during the attempt at killing. Thus a parent, ignorant of his child's biological dependence on the technique of a laboratory employee, risks having him develop the very disease from which he seeks protection.

2 E. B. Lewis, *Science*, CXXV (1957), 965.
3 R. Paschke, R. W. H. Chang, and D. Young, *Science*, CXXV (1957), 881.

A clinical disaster, in the practice of a physician equipped with a sound scientific training, may be the key to a discovery of great importance. The current widespread use of certain sulphonamides as oral substitutes for insulin in the treatment of diabetes mellitus, owes its justification to such a mishap. Janbon, in occupied France in 1942, treated typhoid patients with a new sulphonamide. Some of these developed severe hypoglycaemia and other signs which may follow over-dosage with insulin; a few died. Janbon consulted a physiologist, Loubatières, who spent four years investigating the action of this drug on animals. He concluded that it was "an agent exciting insulin secretion," but continued physiological, pharmacological and toxic investigations for nine years before risking clinical trials. Meantime other investigators who read his reports tested a kindred drug, and in 1955 a group of German clinicians published the results of its successful use in man. If the very large number of patients who are now taking this drug continue to show benefit without harmful side effects, the new treatment will be a great boon and convenience to the diabetic of middle age, but the attitude of the medical scientist is indicated in the statement of Levine, "To me the most important aspect of the research in this field has been the stimulus it has provided to renewed work on the etiology of diabetes mellitus and on the synthesis of insulin, its storage, and the control of its release."[4]

Civilized man looks to medical science to prevent his illness, to keep him fit for employment, to prolong his life. He has good reason for his confidence. We multiply; infantile mortality is low, and the expectation of life for a newborn boy has increased in Canada from 60 years in 1930 to 68 years today. The acute infectious illnesses native to this country have been largely mastered, by vaccination, passive immunity processes, specific forms of treatment, education and public health measures. It is not generally recognized, however, that movement of large numbers of troops or other employees between this country and another has introduced infectious and parasitic agents which are foreign to us, are insidious, and are slowly spreading.

The accent in medical research now is in the field of the degenerative diseases. We shall continue to lengthen the span of life, ignorant of means of maintaining the efficiency of the life we have prolonged. Physical survival to old age is commonly accompanied by infirmity of body, and, more sadly, mental deterioration. Man, interested in his own survival, is not ready to admit, even if he knows, that nature is interested only in the continuity of life, and that he is unimportant in her plan after he has made his contribution to the stream. Through our care of

[4] R. Levine, *Annals of the New York Academy of Science*, LXXI (1957), 291.

the aging, the infirm and the mentally incompetent, we are gradually increasing the percentage of our population who must be dependent on the young and healthy for its food, its shelter and its care. In a country with as much living space as we possess, with such natural resources as confront us, this will be no problem during our life, or even that of our children's children's, but one day, civilized man on this continent will find, as have the races in Asia, that nature will exercise her own methods of control of population.

Man's biological dependence is on himself. The procreative urge will ensure his reproduction. He has learned to adapt his living to his environment, and, to some extent, to harness the natural forces which compose it. He has developed an elaborate civilization, saved himself from his ancestors' diseases, and lengthened his span of life. He has yet to learn to dominate himself, to avoid becoming the slave of his inventions and losing his own initiative and self-reliance. If he should perish, it will be through the uncontrolled destructive powers of his own creations. This would be the ultimate penalty of man's ignorance of his biological dependence.

R. H. MANSKE

WHEN CHEMIST TURNS PHILOSOPHER, he has the advantage that his premises are scientific facts, and since philosophy is one of the branches of learning in which the layman is as authoritative as the professional, I make no apology for this excursion into a foreign field. I begin by pleading guilty to sins of omission but I beg your indulgence because I think that our ethical standards are as high as those of doctors, philosophers, plumbers, lawyers, politicians. Our chief sin has been our timidity. We accept the products of science, but not its ethical philosophy of reasonableness, co-operation, and freedom of thought. We have faith in our theories only so long as they do not run counter to current correct political thinking. Many of you will remember the example of the Scale Boy—that glass tube with a globule of mercury in it—which would not only condition your boiler water and soften your bath water but, used as a swizzle stick, conferred bouquet upon your "old-fashioned." I maintain that there was enough known about the chemistry of water treatment at the time to assure the politicians of the day that the gadget was a fake, and yet a public research institute under the leadership of a military man employed a highly trained chemist to investigate this sheer nonsense. As if adding insult to injury the chemist's final report was never made public. It is true that the Scale Boy went the way of the Great Auk and the Dodo but only after a gullible public had lost its patience along with its money.

I will indulge in the luxury of confession only once more and again I stress a negative. Chemists have spent their time and the public money on work with materials because they are abundant—straw, bitumen, wheat, to name a few. Now I am the first to encourage researches on such problems but if the driving forces are their abundance and hence their economic utilization, I cannot condone such researches. The straw problem of Western Canada was solved by new agricultural practice and not by the chemist and the surplus of wheat is a political matter and the chemist should say so. I am surprised that the honourable member of parliament for the Rocky Mountains has not advocated an all-out research attack on limestone.

I need spend little time reminding you of the positive achievements of chemists. They have been extensive, revolutionary, and often spectacular. We live in an age of synthetic drugs, synthetic fibres, synthetic vitamins, synthetic rubber, synthetic flowers. We have drugs that put us to sleep, drugs that relax us, drugs that alert us. But man is a biological entity and if we tamper with one gland we get unexpected results in another. *In vitro* experiments do not always insure *in vivo* results, because, as long as there is life in the organism, we are dealing with a dynamic system in which the process of change is itself the most fundamental attribute. Though change in a system does not necessarily imply that it is a living one, it is impossible to have a living system without change. If we should ever succeed in getting a complete description of a living body at a moment in time we may learn that time's arrow is missing. We would not be able to ascertain whether the organism is getting old or young and a complete description must include knowledge of the rate of change of entropy and take cognizance of the fact that a living individual has a beginning and an end in time regardless of its reproductive capacity.

What science should strive to achieve is an understanding of the role of each chemical substance as it takes part in the multiple functions of a living cell—how it arises, what it does that cannot be done by something else, and how its role fits into the general dynamic process of life. The brilliant researches of Calvin and his school on the photosynthetic reactions involving carbon dioxide have paved the way to a method of attack on other vital phenomena. The full resources of the organic chemist will be needed to identify complex compounds whose life expectancy is often less than minutes and which never constitute more than infinitesimal amounts. The skills of chemists of many persuasions will also be necessary to unravel the chemistry of enzymes and those proteins which seem to constitute the borderland between animate and inanimate matter. I refer of course to viruses which appear to be a form of life that can be obtained in the crystalline state, and it is by no means impossible that this will be the first form of life to be synthesized in the laboratory.

It will not be many decades before the end products of cell metabolism will have given up their identities. The same is true of those stable intermediates—sugars, fats, some alkaloids, and so on—that have been the concern of the organic chemist for about a century. These constitute either an end or a resting stage in vital processes. The secrets they still hold are their functions, if any, and how they come about. It begins to appear as though genes are merely large molecules of proteins and we may yet learn how a particular sequence of amino acids may influence heredity for good or bad. While it is not difficult to appreciate the survival value of certain mutations it is a little harder to visualize the mechanism whereby such mutations are achieved. The piecemeal removal of amino acids from a protein molecule and the generation of a new gene are problems that await new tools. We may first have to answer the question why species are as stable as they appear to be.

I have not mentioned inorganic chemistry, the reason being that when inorganic elements take part in vital processes, they come into the domain of the organic chemist. I suspect that the role of the so-called trace elements will be unveiled in the near future. Whether they will all end up in stable organic compounds like cobalt in vitamin B_{12} remains to be determined.

The role of physical chemistry in life processes has long been known to be an important one and physical chemists can make crucial contributions to such processes by learning more about the behaviour of high energy organo-phosphorus and other compounds.

I would finally like to enter a plea that scientists take a more active part in public affairs. Their whole training has been designed to fit them for objective thinking and if such a habit can be employed in the solution of problems of public affairs, it is likely that fewer gold bricks will be bought by the public. Let us make a mockery of Robert Sherwood's statement that "nature has taken the world away from the intellectuals and given it back to the apes."

THE SOCIAL IMPACT OF MODERN TECHNOLOGY

N. A. M. MacKENZIE, F.R.S.C., *Chairman*
V. W. BLADEN, F.R.S.C.—E.W.R. STEACIE, F.R.S.C.
W. H. WATSON, F.R.S.C.

N. A. M. MacKENZIE

BECAUSE THERE ARE a number of others on this panel who have given time and thought to the papers they will present, my introductory remarks will be brief. But I do want to join with others in expressing my thanks and congratulations to Dr. Watson and the members of his committee who have planned and organized this Symposium. I have long felt that the members of the various Sections of the Royal Society should meet together as a group more often, not only to listen to speeches and to eat dinners, but to discuss with others in other disciplines and walks of life matters that are important and of interest to all of us.

The topic for this session, the impact of modern technology upon our society and upon the future of our nation, seems to me at least to be one of the most important that the Royal Society and other learned societies in Canada might consider. There is no doubt whatever that the kind of society that science has made possible is going to create major changes in human life. The impact of the concentration of population in great cities on the problem of transportation, as well as the speed and ease of transportation, together with the changes in our physical environment which the scientists will no doubt tell us about later on— all these make it certain that the lives of our children and of our grandchildren will be very different from the life that we have been familiar with and that we took over, as it were, from those who preceded us.

V. W. BLADEN

THE SELECTION OF THIS TOPIC, "Our Debt to the Future," as the subject for discussion at the seventy-fifth annual meeting of this Society may be, and I would like to believe is, a symptom of an important change in the climate of opinion. The appointment and activity of the Royal Commission on Canada's Economic Prospects may be considered a more important, more significant, symptom of the same change. Consider the climate of opinion of the thirties: the "stagnation thesis"; the expectation of declining population; thrift as the serpent in the economic Eden; the too ready acceptance of the cynical Keynesian quip, "in the long run we are all dead." No wonder Harold Innis had so much to say in the forties about "time," about the "foreshortening of practical prevision," about the fatal obsession with things of the present. Perhaps I am unduly optimistic. The climate of opinion has changed under the impact of changed conditions, a resurgence of invention, investment and population, in spite of a continuing threat of a war which, if it comes, may well be a war to end wars in a sense not contemplated when that phrase was coined. May one hope that this new era of growth may liberate the human spirit as earlier ones seem to have done. May I mix with the pessimism of those who fear an age of materialism, luxury, and possibly destruction, some of the optimism of David Hume writing in 1752.

In times when industry and the arts flourish, men are kept in perpetual occupation, and enjoy, as their reward, the occupation itself, as well as those pleasures which are the fruit of their labour. The mind acquires new vigour; enlarges its powers and faculties . . .

Another advantage of industry and of refinements in the mechanical arts, is that they commonly produce some refinements in the liberal. . . . The spirit of the age affects all the arts: and the minds of men, being once roused from this lethargy, and put into a fermentation, turn themselves on all sides, and carry improvements into every art and science. Profound ignorance is totally banished, and men enjoy the privilege of rational creatures, to think as well as to act, to cultivate the pleasures of the mind as well as those of the body . . .

The more these refined arts advance, the more sociable men become . . . they flock into cities: love to receive and communicate knowledge; to show their wit or their breeding; their taste in conversation or living, in clothes or furniture. Curiosity allures the wise; vanity the foolish; and pleasure both.

That there is another side to the coin of economic growth, I do not dispute; on that side we would see depicted misery, cruelty and ugliness. For a description of this side we can turn to Karl Marx or to the Hammonds. But history has not entirely falsified the hope of Hume and need not entirely falsify our hope of human improvement. Cliffe Leslie writing in 1862 remarked that the change in the "quality" of wealth was more remarkable than the change in its "quantity." New desires for health, decency, knowledge, refinement and intellectual pleasures have, in fact, revolutionized production. And Dr. Dorothy George has shown in her *London Life in the Eighteenth Century* that there was a marked improvement in the character of the working class from 1780 to 1820. We ask a lot, however, when we ask improvement of the people, not just improvement of an *élite* supported by a brutish mass.

Along with optimism too often goes complacency. The Gordon Report promises "that by 1980 the average Canadian after paying income tax will have about two-thirds again as much net income for his own use as he had in 1955," and this with an average working week of 34.3 hours. The danger is that we may forget our debt to the future; that we may assume that the fruit will drop in our hands without effort; that progress towards this potential of 1980 is mechanical and inevitable and independent of our current behaviour. It is an obligation of social scientists to consider, and to discuss in public, the conditions under which we are most likely to enjoy this increase in material production. But at the point where I make some brief comments on this problem, let me issue a warning against the search for simple solutions. Let us take seriously to heart the warning given by Dr. Innis in his essay on "Discussion in the Social Sciences" (*Dalhousie Review,* 1936): "As an economist I am constantly faced with the extraordinary difficulty and complexity of the social sciences and constantly forced to admit defeat. If an economist becomes certain of the solution of any problem he can be equally certain that his solution is wrong." As I said in a paper presented to Section II last year, the economist must be modest, recognizing the element of art, the extent of ignorance, the inevitability of uncertainty. "Yet this modesty must not degenerate into a timid scepticism, his avoidance of the doctrinaire into simple indecision."

The problem before us can be stated as the identification of the conditions under which we might maximize the net advantage to humanity from the advance in science and technology. This involves, first, the conditions most favourable to increase in material production; second, judgment in the choice of goods to be produced (remembering always that leisure is a most important "good"); third, recognition of the human costs of rapid change and consideration of the means of minimizing such costs. We must note that any reduction in the cost of change not only increases the net advantage of any given change, but also increases the speed and magnitude of change by reducing opposition to it.

Of the conditions favourable to increase in production, perhaps the most important are those under which pure science flourishes, under which the application of science to industry is encouraged, and under which an adequate supply of men with the requisite skills to utilize the new scientific techniques in industry is assured. This is not primarily a matter of economics, though a generous financial provision for education and research is not unimportant. Two conditions seem to me to be important. The first is academic freedom in the fullest sense. The second is the revival of the aristocratic tradition in education. Democracy was right to revolt against educational privilege for the old aristocracy or for the new plutocracy. It was wrong in not providing the best possible education for the aristocracy of talent. It failed to realize that such provision was not a matter of individual privilege and right, but of social advantage, even of social necessity. Again, I turn to Harold Innis: "A democratic society can thrive only by the persistent search for its greatest asset, and by constant efforts to conserve, to encourage, to train, and to extend it. . . . Universities must strive to enlist most active energetic minds to train most active energetic minds." (*The Bias of Communication*, pp. 207–9). Yet we hear professors of education denouncing special classes for the gifted, preferring the well adjusted to the brilliantly creative! Unless we imbue our children with the love of excellence, and give them the joy of maximum achievement, we cannot expect to reach the high levels of production that are within our grasp, we can scarcely hope to survive as a nation in an insecure world, and, what is much more important, we shall have failed to give them individually the key to a greater happiness than is open to well-adjusted mediocrity. The human ideal cannot be that of the "contented cow."

The economist must concern himself particularly with the conditions favourable to business enterprise, to industrial expansion. Professor Keirstead will have something to say about the provision of the necessary capital; but first there must be the will to invest, the will to venture. The risks are great. As Professor Schumpeter said in his stimulating book, *Capitalism, Socialism, and Democracy*: "Long range investing under rapidly changing conditions, especially under conditions that change under the impact of new commodities and new technologies, is like shooting at a target that is not only indistinct, but moving—and moving jerkily at that." It is important that there shall be pressure to invest and to innovate, and this comes from competition. Seeking security through cartels or restrictive agreements of one sort or another, business men might sabotage progress. Hence the importance of our legislation against "combines." But no pressure to expand can be effective if the risk is too great. Some mitigation of competition in the short run, I have argued before the Gordon Commission, is necessary. This is the old dilemma of security and progress: too much security and too little security are alike unfavourable to progress. Two favourable factors may be noted. First, risks of expansion are much less when population is growing. Second, risks appear smaller, the higher the level of employment that is expected. (I would add that full employment seems to me to be a most effective solvent of restrictive agreements.) Much, then, depends on the succcess of our full employment policies. We have learnt a lot since the thirties; only time will tell whether we have learnt enough, or have learnt the right things. One problem is already insistent. Can we have full employment without inflation? Perhaps more important is the further question: whether inflation is as inexpedient as it is unjust. Most social scientists have been victims of its injustice and may be suspected of bias in estimating its expediency.

Let me close this section with a plea for research and reflection on the conditions favourable to "investment." The influence of Keynes has led too many to treat the volume of private investment as independent of social policy, and to concentrate on fiscal and monetary policies devised to maintain stability in face of the fluctuation, and the expected inadequacy, of such investment. In the thirties, I preached the importance of investigating the determinants of investment: in the fifties, when investment is more than adequate, I suggest that such investigation is still important lest we inadvertently stifle our present expansiveness, or fail to take appropriate measures if the present rate of investment should unduly decline.

I turn next from quantity to quality, or from quantities of things to quantities of "utility," or satisfaction, whatever that may be. When the Gordon Commission promises the average Canadian 66 per cent higher net income, this means that much more money to spend. But the things he will buy will be different and to compare the relative magnitude of his "real" income then and now involves the familiar difficulty of adding elephants and giraffes. However, we may accept the change in money income over this period as a useful index of the change in real income. Net human advantage is a matter of "utility," of satisfaction, of "psychic" income. The psychic income from different things may be homogeneous, and might be capable of being added were it capable of being measured, which it is not. So at this point, I leave problems of quantity and suggest some problems of quality. And first I return to the importance, noted by Cliffe Leslie, of improvement in the quality of our wants. I shall not seek to impose my own scale of values, in which, for instance, television ranks very low. Nor shall I preach the doctrine of asceticism, the road to plenty through reduction of wants rather than through provision for their satisfaction. But I would like to suggest that a most important change from this point of view would be reduction in the intensity of that "competitive" consumption of which

Veblen wrote so effectively. That the satisfaction of old wants will lead to the development of new ones can be accepted as healthy; but that satisfaction shall not increase because the more the Joneses have the more we must have to remain equally satisfied, involves us in an endless pursuit of an unattainable and futile end.

We are promised the higher net income along with a shorter work week; which leads me to say something about leisure. In the Annual Report of the Twentieth Century Fund for 1956, Dr. August Heckscher discusses this new "problem in American values":

> Next to the abundance of things the citizen has, the abundance of time at his disposal is perhaps the most striking characteristic of the present American scene. . . . Leisure time has before this belonged to a small group in society; it has been won at the cost of slavery or, at best, of toil and drudgery by the great mass. But now leisure is promised for all.
>
> Free time is not in itself leisure . . . to describe mere idleness and self indulgence as leisure is to use the word in a questionable sense. The classes, which in great historical periods maintained learning and manners, which pushed back the frontiers of the physical and intellectual world, have made leisure synonymous with freedom, and freedom synonymous with self-discipline and disinterested effort.

The problem is to ensure that the free time that we are promised may be transformed into leisure. This raises many new problems; not only problems of education for leisure but problems of organization of free time. The distribution of free time over the life span raises questions about our current retirement practices. Dr. Heckscher asks whether a "three day dosage at the week's end will eventually prove as useful for leisure as continuance of the present week end with a free day added somewhere toward the middle of the week." And I have speculated, in a short paper read to the Industrial Relations Research Association in December 1956, on some of the problems involved if we all want to enjoy our free time at the same time as others.

Let me next consider the possibility of increasing the net human advantage from production by reducing human cost. Such inquiry is very much in the tradition of the classical economics, from the "real price" of Adam Smith to the "real cost" of Alfred Marshall. Neo-classical economics turned away from these considerations of human cost as it became obsessed with problems of equilibrium and allocation of resources. The great heretic, J. A. Hobson, developed the theme of human cost, but he made little impression and few read him these days. But as we become more concerned with wealth than with equilibrium, with growth than with static models, we should return to a consideration of human cost. Of course, we will be dealing with quantities that we cannot measure, with problems where subjective judgment cannot be submitted to objective testing. Yet there can probably be agreement on some general propositions. Whereas John Stuart Mill in 1848 doubted whether machines had lightened the work of any human being, it is surely clear that modern machine industry has reduced, and will still further reduce, the strain on human muscle. The problem now, and for the next generation, is to reduce the psychological strain.

Here an economist is quite out of his depth: but it appears that the researches of Elton Mayo and many others have demonstrated that slight changes in the social organization of the factory, or the office, can lead to great increases in productivity and decreases in fatigue and mental stress. Mayo's experiment in a Philadelphia textile mill is the classic case. Later studies have indicated that one of the serious costs of technological change is that social relationships are destroyed and personal stress results. When more is understood about human behaviour in such situations, this cost can be reduced, opposition to change can be reduced, and we can enjoy the increase in production (or the increase in leisure) that such change makes possible. There are dangers: charlatans are rife, and malignant use can be made of such knowledge to manipulate men as in the Orwellian vision. But to my mind research in this field, and joint study by scholars, business men and trade union leaders, are urgently needed. It is a matter of regret that the programme of research in this general area which was carried out in the University of Toronto by the Institute of Business Administration and the Department of Psychiatry has petered out. I believe, and I have argued elsewhere, that not only the success of our economy but the future of liberal society depends largely on our success in finding ways of making efficient production consistent with the "good life."

Resistance to change is not only a matter of psychological insecurity. It is natural that workers should seek to resist those changes that threaten their individual earning power; though their success may reduce the standard of living ultimately attained by the workers as a whole, or by the children of these individuals. Few investments, I venture to suggest, would be more productive in the long run than generous compensation to skilled workers for damage to their earning power resulting from technological improvement, and generous and imaginative schemes for retraining to enable the skilled whose skill has become obsolete to acquire new skills.

Finally, let me remind you that our debt to the future is not simply a debt to the future of Canada. We are citizens of the world: a world which, however divided politically and ideologically, is still "one world."

Perhaps our purely selfish concern for the future of Canada imposes on us a concern for the future of the rest of the world: certainly to most Canadians such concern is accepted as a duty (though lip service is as yet more evident than action). Perhaps I can make my point best by referring to the little book *The Challenge of Change* in which Lawrence Thompson gives a preliminary report of the Duke of Edinburgh's Study Conference at Oxford last summer. In a Commonwealth setting the world problem emerges. We cannot ignore it.

E. W. R. STEACIE

IT SEEMS TO ME that I am placed in a somewhat difficult position since the subject for this session is "social" and my colleagues are professionals in the social science field. Presumably I am tossed into this situation to illustrate the disadvantages of a narrow, specialized education. In any case, it would, I think, be inappropriate for me to attempt to deal directly with the subject of "The Social Impact of Modern Technology" in such company. I propose therefore merely to point out a few aspects of the situation, as they appear from the other side of the fence, and which are frequently confused or ignored.

Of the definitions of technology which are given dictionary recognition, I think the one which is implied in this afternoon's discussion is "the practical arts considered collectively." If one accepts this definition it seems to me that a great deal of discussion on the social impact of technology is rather inept. The question is frequently asked "How does technology affect society?" in somewhat the same way that one might ask "How does measles affect society?"—in other words as though technology was a quite extraneous influence. Now, in fact, society and technology involve the same people and the same things, in the sense that technology is merely the sum total of what everyone, or almost everyone, does for a living. (There are, of course, a few exceptions, humanists, pure scientists, artists, and so on, but they do not constitute a large percentage of the total population.)

It should also be emphasized that this has always been the case even in preliterate societies, or perhaps one should say especially in preliterate societies. The impact of technology on society is therefore merely the impact of what society does upon itself. It is by no means an outside, unpleasant force exerted on society by a few engineers, but is the collective influence of everyone's actions. The salesman is just as responsible as the engineer; in fact even the humanist gets into the game via T.V. commercials!

The real problem is, of course, not technology itself but rather technological innovation: this is what upsets the peaceful course of our lives. Technological innovation has always been with us. The problem in recent years has mainly been not the increase of technology, but the rapid rise in the rate of technological innovation. It often seems to be suggested that such technological innovation is a juggernaut which rolls along crushing society in its course, and that society has no power to combat or modify its effects. This is the exact opposite of the true situation.

Science has developed an increasing understanding of nature. As this knowledge and understanding develop there is an increase in the pool of natural knowledge on which technology is based. The technological innovation that results, that is, what is invented, is then a matter for society to decide. Far from technology forcing itself on society, it is society which ultimately controls technological innovation. In other words society gets the inventions it wants: these of course are not necessarily what it needs.

As a result it is significant that some of the most far-reaching inventions from a social point of view are based on quite trivial scientific advances. The outstanding example of this is the development of the internal combustion engine and hence of the automobile. It had been known for centuries that gas mixtures explode, and that explosions will push things. To make them push a piston was certainly not revolutionary. Even the mechanical ingenuity required was not of the highest order. Yet the automobile produced the most profound social changes. Surely the reason for this was not that society was forced to use automobiles, but rather that the automobile was wanted, and was therefore invented.

A given technological innovation is therefore by no means inevitable, but is a definite and deliberate choice of society: whether society exercises this choice in a sensible way is, of course, quite another matter. Certainly no one makes a gadget for long if he can't sell it. Hence out of all the innovations possible at a given moment, on the basis of the existing store of scientific knowledge, the ones that are chosen are the most wanted. (They may not, of course, be the most needed.)

The main point in all this is that the argument is often made that technological advance has a great influence on society, with the tacit implication that society has little influence on the direction of technological advance. Actually the direction of technological advance is apt to be due far more to advertising and sales and promotional efforts than to the efforts of scientists and engineers. For example, it is equally possible from a technical point of view to have automobiles get longer, more ornate, higher powered, more automatic, and more expensive to buy, operate and repair or to have them get more durable, cheaper and more convenient. The direction of the development is decided by the public under the influence of mass media of communication. There is no question therefore that the promoters have more influence on the direction of advance than do engineers and scientists. In fact there is every indication that the direction of this technological innovation is decided much more by the possessors of B.A.'s than by those with B.Eng.'s and B.Sc.'s. A broad general education seems to have been rather wasted on those in the advertising business.

Incidentally, one sometimes hears expressions of worry about the unplanned nature of technological innovation. This is usually coupled with the suggestion that there is far too much delay between the accumulation of scientific knowledge and its application to technological development. It is suggested that a fully planned technology would speed things up enormously. In part this may be true, but there are two other aspects of the situation. In the first place there is little doubt that the hurried over-planning of such development would inevitably result in so much confusion and report writing and so many committees that the whole thing would be slowed down instead. There is also the other side of the picture: ignorant exploitation of science at least puts the brakes on. A fully effective planned application of science to develop everything under the sun as quickly as possible is too horrible to contemplate. There is only so much technological innovation which any of us can stand at any one time. In fact there is nothing more distressing than the predictions of enthusiasts as to how automatic our lives are going to be in a hundred years. Some time we will have to make up our minds as to how much efficiency we really want, and in what direction we really wish to proceed, efficiently or otherwise.

The crux of the matter is that the development of scientific knowledge, and the potential technological advances which may arise from it, have given society the chance, for the first time, to make decisions on many matters which in the past had been largely or totally beyond its control. For example, in the past the population of the earth or of any given part of it has been largely dependent on disease, the birth rate, and so on. Today, for the first time we have the information, the control of disease, etc. which would enable us to make effective decisions about population. There thus arises the question whether society is willing to make any decisions at all about the matter, as well as how intelligent such decisions might be. In the meantime, of course, the decision to do nothing is just as definite a decision as any other, and the consequences of such a decision cannot be avoided. It is, however, essential to realize that potential technological innovation is offering society freedom and not the reverse. At the same time it is making it essential for society to seize the opportunity to make decisions, and the future will bring up many awkward questions.

It is perhaps worth emphasizing also that there is a great deal of loose thinking on the question of the moral responsibility of science and scientists for things like nuclear weapons. All science can do is to increase the fund of natural knowledge and thus increase our potential control over our environment. What society does with this power is a social problem. There is no advance in the arts that cannot be used for good as well as for objectionable purposes. If writing had never developed there would be no yellow journalism and no comic books, but I doubt if my humanist friends would agree that the development of writing was unfortunate.

I was told by the Chairman that I should attempt to be provocative, and in obedience to this instruction I would like to sum up in the following terms: If society is going to the dogs because of the misuse of technological innovation, and the inability to make correct or suitable decisions when the need arises, it is not the fault of the scientist who has given society an increasing supply of information on which to base its decisions. It is, rather, the fault of the humanists, the social scientists and the theologians who have been unable to cope with the problems which have been raised by our increased knowledge of our environment.

If there is a conflict between the scientist and the humanist today, which I doubt, it is not a problem of "narrowness" *versus* "breadth," but rather one of the relative simplicity of understanding nature as compared with understanding man.

W. H. WATSON

IT IS VERY DIFFICULT to grasp what is going on in technology without experience as an engineer or a scientist. Very dramatic changes are taking place in the United States today, and it is hard to believe that in Canada we shall not see corresponding developments. The public has learned of the use of electronic machines for accounting and other functions in business. Automation in factories dependent on electronic computers has become the fashionable theme for executive vice-presidents and trade union leaders. But the most important role for computing machines in the future of our country is hardly mentioned. It is to aid the engineering designers and scientists. Through them, we should expect the mathematician to come to play a far larger expert role than he has ever done before.

When an aeronautical engineer is designing a wing for an aircraft, he is concerned with the deformation of the wing by the forces on it in flight. These forces are caused by the air flowing around the wing, and the pattern of this flow is determined by the shape of the wing. But of course this shape is determined by the forces due to the flow and those called into play by the deformation of the wing. The internal structure of that member is a complicated affair of struts and ties. Without a high-speed computing machine, the detailed solution of this problem in design was unattainable. With its aid, much of the expensive trial-and-error experimentation with models can be avoided. The result is that aerodynamic design has much of the guesswork taken from it. This is a basic aim in engineering—to take the guesswork out of construction and out of the operation of machines.

In like fashion, the design of bridges, of atomic reactors and of hydraulic systems in power generation, to mention three notable examples, is aided with an efficacy that parallels that conferred on locomotion by the invention of the internal combustion engine. Insofar as engineering design is concerned, the engineers of the future will be as far ahead of those of 1950 as the user of power tools is ahead of the hand craftsman. More attention will then be devoted to the theoretical conceptions embodied in the design. The designer will have greater freedom to explore novel ideas. The amount of detail that can be scrupulously examined will be limited mainly by the time required to obtain it, for the cost will be trivial compared with that required to obtain the same information in the old way. The engineers and mathematicians employed on such work in the future will engage in imaginative investigation, experiment with ideas, in short cultivate the mental attitudes of the experimental scientist in the context of problems of practical importance. There are many old engineering procedures in daily use that call for imaginative criticism by intelligent men untrammelled by hoary tradition and unimaginative practice.

I have presented, very briefly, as I see it, the spirit of modern technology and I do not doubt that young Canadians attending public school today will experience many of the practical effects of the new methods in engineering design before they have families of their own. If we do not develop our own systems for doing these things ourselves, they will be done for us in the United States at our expense. For it is quite certain that the new methods of design will be superior in every way to conventional methods of the past and, in the long run, much less costly. This will cause designs to be purchased in the United States, if our engineers have not the training and facilities to do it themselves. It is clear that much more is at stake than technological specialization. On our ability to imagine what is needed and to act intelligently in anticipation of the future in applied science and engineering may depend, in great measure, our political independence on the North American continent.

You will observe that I use the word "technology" in the sense of "applied science" and practical knowledge, without implying that it is a lower form of knowledge. To my mind, technology requires the best brains we get, stimulated by the ablest teachers we can find. It is not a sort of sub-human activity suitably performed by slaves or dullards, for technology is perhaps the most effective cause of social change today.

In spite of the common opinion that technology dehumanizes men, and the academic opinion that to attempt something practical requires a lower grade of intelligence than is required of the writers and talkers, the facts of life with respect to the growing invasion of life by inventions will require on materialistic grounds alone that men value competence and integrity, not at the say-so level of public report, but in the recesses of their hearts and minds. Instruction in practical matters has indeed great practical educational value in developing a sense of moral responsibility, provided that teachers insist on comparing achievement by the pupils with what they set out to do. It is because technology faces the problems of making things work or getting practical answers that objective tests of performance are inherent in its operation. I predict therefore that the wider use of machines will require more from the men whose work is associated with them. A fast computing machine that is capable of supplying 10,000 correct answers per hour is capable also of providing 10,000 wrong answers per hour if someone has erred in preparing the work for it.

We have to learn to combine the solid tradition of broad human understanding with an appreciation of the need for new intellectual tools adequate to deal with practical matters. Far too frequently, in recent years, professional educators and others have made public speeches that appeared to say something about the needs of society. Too much emphasis, they say, has been laid on technical instruction, and real education is neglected. But it is fairly obvious that their conception of real education is an ideal from a former generation, free from disturbing thoughts of the intellectual challenge of a rapidly transforming society. It is this acceleration of the pace of social change produced by science and engineering, and in the course of the natural development initiated by the great social reformers last century, that we must mark. The great fact in Western society today is that the electorate is not always ready to accept and tolerate social conditions that in previous generations were regarded as inevitable. The ordinary man in our time knows a great deal more about how the world works than did his forebears. We are only now seeing the results of mass education as an influence accelerating social change.

I emphasize the word "accelerating." We have inherited concepts of a static, stable society. The first industrial revolution compelled man's attention to change and rate of change. I submit that in this revolution that is now upon us, we must think of the next higher derivative, because the changes taking place in our generation are so striking. Instead of employing merely the simple categories "changing" and "unchanging," we should always keep in mind the time scale of their operation. Processes that take place slowly may be treated as if they produce no change at all over a period of time that is brief compared with that required for significant change under them. Causes that produce catastrophic effects must be considered in relation to the frequency of their occurrence, in order to estimate the long-term effects due to them. More important still, we must learn to distinguish divergent from convergent processes. The latter lend themselves to management by the normal processes of government, provided that the time during which the change takes place is not too short, nor the magnitude of the change too great.

A divergent process, on the other hand, becomes more difficult to control, the longer one waits to act. For example, any physical chain reaction is easily controlled in its early stages, but, once it has grown to substantial proportions, quite exceptional efforts may be required to achieve control, if that is possible at all. Human experience with fire is the best testimony to this. But other impressive examples can be chosen from engineering with chemicals or nuclear material. To what extent these conceptions can be applied to society has not really been investigated on a scale large enough to find out. The study of the course of epidemics may have much to teach us about the mechanics of some of the multiplicative processes by which the course of orderly social evolution may be catastrophically diverted.

Are we on the steeply rising curve of runaway social change? Does the evolution of social change resemble the course of a divergent nuclear reaction? If so, where are the control rods to make this process convergent? I believe that limits to the rate of development and transformation of society will arise naturally in man's biological make-up.

What is the point of accelerated social transformation? One might just as well ask the question "What is the point of the growth of a bacterial culture in a nutrient medium?" This is the law of its nature. The evolution of science—as of other human activities—is a natural phenomenon. Man's conscious attempts to control these activities are also a part of nature—an extrapolation of the complexity of biological adaptation to our environment. We are in process of evolving now new ways of working on our environment to change it for many varied purposes. But we do not have yet an effective balancing mechanism of over-all social control resembling the integrating action of the central nervous system, making every part of our bodies in health serve in its own particular way the needs of the body as a whole.

The effects on the individual person that could develop from the invention of such mechanism should surely be viewed with concern. Technology is putting man on the spot, morally. For me, this is the most important aspect of its social impact. There is little practical value in viewing with alarm. A very great deal of re-thinking is required in universities about these problems, and the sooner it is recognized that intellectual stimulus for our time lies in the sphere of the technologists, the sooner we will give up play-acting about what constitutes effective thinking.

The attitude to science and to our rapidly evolving technology on the part of some individuals who think they are thereby defending the humanist position is to insinuate—when they do not openly avow—a contempt for man's mechanical inventive genius, as if man, exercising one of his greatest powers, must, in the process, be de-humanized. Here is a statement made recently in a symposium on the automatic age: "In building a world where machines do the work that used to be done by men, it is not good enough to build men who can work only with machines. It is not enough to produce managers who know how to increase production; of far greater importance is it to produce managers who know why they should increase production." This is a fair example of the kind of superior comment on the present situation that we can expect to hear reiterated so often that we shall come to ignore even its lack of logic. Of what practical value is it if one understands why or whether one should increase production if one does not know how to do it? Surely, one should say that because we now have the means to increase production to a noteworthy degree, the questions why and how now have some point.

In the university, one deals really with imagined problems and discusses the principles and techniques for treating them. Actual problems are far more complicated than these imagined ones. Their solution is usually made more difficult by the technical inadequacy of the means available. Modern technology is effective in providing adequate means because of scientific research. A common academic attitude to this rapid development of science and technology is to question its intellectual, moral and spiritual importance; "mere knowledge," it is said. But only persons who do not, in their experience, find the need to use this knowledge will acquiesce in this evaluation. The socially important fact is that many of the men who by endowment, training and experience are entrusted with commenting on the broad human situation are often unable to comprehend the scope of the changes that are imminent. By separating themselves from science, they have lost touch with a large and important part of the intellectual life of modern man. The methods of science will invade any intellectual territory if the appropriate tools can be found to manage the information pertinent to it. This is not to say that the methods of science are the only methods that men may be disposed to use, but whenever these methods make possible for man the intellectual equivalent of the combine in agriculture, primitive procedures will go out of use. This is one important aspect of the social impact of modern technology because, through efficient educational processes, new ideas work their way with cumulative effect to cause social change beyond our wildest imagining.

OUR ECONOMIC POTENTIAL
IN THE LIGHT OF SCIENCE

H. C. GUNNING, F.R.S.C., *Chairman*
J. E. HAWLEY, F.R.S.C.—L. M. PIDGEON, F.R.S.C.
B. S. KEIRSTEAD, F.R.S.C.—MAURICE LAMONTAGNE, F.R.S.C.

H. C. GUNNING

THE SUBJECT of this part of the symposium is "Our Economic Potential in the Light of Science." The topics for discussion will include: non-renewable resources—energy sources, metals, and industrial minerals; conservation and replacement; the trend towards increased use of metals (in addition to iron), which are abundant in the earth's crust but require large amounts of energy to produce; the economics of the conservation of resources and economic factors in technological progress.

Your committee has assembled an outstanding panel, as you see: one geologist, one metallurgist, and two economists. Dr. Hawley is Miller Memorial Research Professor and Chairman, Graduate Studies, Department of Geological Sciences, Queen's University. Dr. L. M. Pidgeon is head of the Department of Metallurgical Engineering, University of Toronto. Dr. Maurice Lamontagne is economic adviser to the Privy Council of Canada, and Dr. Keirstead is professor of political economy at the University of Toronto.

The discussion will be opened by Dr. Hawley and continued by Dr. Pidgeon, so that the two people who are concerned with our mineral resources will explain that aspect first and make some suggestions about the abundance of mineral resources and their use; after that, the economists will discuss the matter of conservation and comment upon the attitude of the previous speakers.

J. E. HAWLEY

CONSIDERATION OF OUR NATURAL RESOURCES, their development and conservation, in the context of the symposium, "Our Debt to the Future" should involve also, I contend, our debt to the present free world. Discussion, perforce, will be limited to our exhaustible, non-

renewable, mineral resources—fuels, metals and industrial minerals—
as they present the greater problem, compared with renewable, or con-
servable resources, such as soils, water, forests and wild life. Behind
what is said is the thought and hope that the wise use of our material
wealth will result from careful planning over the long term, and that
future generations of Canadians will reap the benefits, rather than inherit
a land from which all the good has been drained.

Studies of this type are currently the fashion in many countries, and
a vast literature[1] is or soon will be available, covering the field much
more completely than is possible here. As Canadians are among the
most mineral-conscious peoples of the world, it should not be necessary
to state in too great detail the facts by which Canada holds high rank
as a mineral producer, nor to indicate why she is regarded as one of the
few remaining great store houses of future supplies. These are the facts
and reasons bearing on our economic potential.

How far we should look into the future has been left to the individual.
Here we need not think in units of geological time, enlightening as this
might be. We must and should, however, look beyond such short periods
as are commonly in vogue.

To make a complete inventory of our known and hidden mineral
resources is an impossible task. No apology need be offered for such a
statement. It is simply an acknowledgment of the limiting factors imposed
by nature. What *can* be done is to indicate our currently known reserves,
and the geological probability of replacing them with others, and to
consider their life in terms of present and future rates of depletion. Even
this requires an understanding of their character, origin and distribution.

Mineral deposits, in the broad sense, come in a myriad of assorted
sizes, shapes and constituents. Few are exposed at the surface; many lie
under a cover of water, soil or rock; several dip outward from the continent, beneath the sea. They have been formed throughout the ages, not
in a day. Their history may be long and complex. Some, the simple,
bedded type, are readily measurable. Others are irregular in form,
occurring as pipes, pods, and lenses, and are more difficult to find and
outline. Some are shallow; others extend to great depths, though only
minable to a maximum of about 12,000 feet, so that economic considerations alone demand that exploration and the delineation of reserves

[1] *Resources for Freedom*: Report by the *President's Materials Policy Commission* (Washington, D.C., 1952), I, *Foundations for Growth and Security* (Paley
Report); Institution of Mining and Metallurgy, *Mineral Resources Policy: A
Symposium* (London, 1956). Royal Commission on Canada's Economic Prospects, *Preliminary Report* (Dec. 1956).

be timed only slightly in advance of actual mining. Some are reasonably uniform in quality, others vary widely, both vertically and laterally. Once deposits are found, these features are determined, but other, less controllable, factors also affect their worth, and require long-term predictions as to future metal values, costs of labour and equipment, domestic and world demand, new uses, possible substitutes, effects of improved methods of mining and extraction, as well as inevitable taxes and possible subsidies. These may vary from year to year and so determine that what is useless today is valuable tomorrow, or, indeed, the converse.

As further background two other points should be made. The first deals with the average composition of the outer part of the earth's surface.[2] Table I shows that only ten elements make up over 99 per

TABLE I

PERCENTAGE ELEMENTS IN THE EARTH'S CRUST

Oxygen	46.60
Silicon	27.72
Aluminum	8.13
Iron	5.00
Magnesium	2.09
Calcium	3.63
Sodium	2.83
Potassium	2.59
Titanium	0.44
Manganese	0.10
	99.13

cent. These are average figures for the composition of the common rocks, some of which contain more, some less of the elements listed. Oxygen is combined with all the others; silicon is of limited use and is usually a hindrance rather than a help. Most of the aluminum, iron, magnesium and titanium are present as silicates, some iron as oxides. In no sense do such occurrences of metals represent profitable or economic ore deposits. Rather may they constitute an ultimate source of such metals when the world's richer, more concentrated deposits are exhausted. They will never be cheap.

Other metals which we consider important, are present (Table II) only in parts per million, but they *are* there, and with man's ingenuity many may be recovered eventually and put to use.

[2] Brian Mason, *Principles of Geochemistry* (New York, 1952).

TABLE II

LESS ABUNDANT ELEMENTS
IN EARTH'S CRUST

Element	Parts per million
Manganese	1000
Chromium	200
Zinc	132
Nickel	80
Copper	70
Tin	40
Cobalt	20
Lead	16
Uranium	4
Silver	0.1
Palladium	0.01
Platinum	0.005
Gold	0.005

A second point is the general history of mineral development in major industrial countries.[3] Here, in contrast to the preceding discussion, we are dealing chiefly with mineral resources in which specific elements or compounds are present in relatively concentrated form (in hundreds or even thousands times the amount present in average rocks), deposits which under existing economic conditions have been worked at a profit. Certain features of this history are of interest to us. Figure 1 (modified after Lovering especially for early Canadian imports and exports), shows a well-defined pattern and five general stages: stage one, a period of

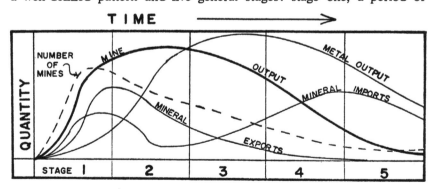

MINERAL AND METAL PRODUCTION FOR AN INDUSTRIAL COUNTRY

(Modified for early Canadian Imports and Exports, after Lovering)

FIGURE 1.

[3] T. S. Lovering, *Minerals in World Affairs* (New York, 1943).

increasing discoveries, development of mines, increasing mineral production and export; stage two, a period of smelter development, increasing output from larger mines, exhaustion of the smaller, and decreasing exports, as stage three is reached with greater industrial development, rapid accumulation of wealth and expanding internal and external markets. In stage four comes the more "rapid depletion of cheap raw materials," a drop in mine output, and an increase in imports. In stage five comes still greater dependence on foreign supplies and competition for their control and profit. Canada is clearly still in stage one, but the shape of things to come is clear. The United States seems to be entering stage four, but already is looking ahead, while Britain is now in stage five, according to Lovering. The present great American investment in our natural resources is clearly not a mere coincidence.

With this introduction let us deal first with what is known, then in probabilities.

Our Mineral Production

Present mineral production[4] indicates the rate of depletion. None the less it is natural not to deplore, but to boast of the imposing record of our mineral industry, recognizing, as we must, the important part it now plays in the economy. First place in the production of five, and from

TABLE III

CANADA'S WORLD RANK IN MINERAL PRODUCTION

1st place	Nickel	80 per cent
	Platinum metals	40 per cent
	Asbestos	60 per cent
	Calcium	
	Nepheline syenite	
	(soon – Uranium)	
2nd place	Aluminum metal	
	Cadmium	
	Cobalt	
	Gold	12.5 per cent
	Magnesium	
	Selenium	
	Titanium (ore)	
	Zinc	14 per cent
3rd place	Silver	
	Molybdenum	
	Barite	
4th–5th place	Copper	9.6 per cent
	Iron	
	Lead	9 per cent

[4] H. McLeod, "The National Mineral Results for 1956," *Canadian Mining Journal* LXXVIII (1957), pp. 75-89.

second to fifth in at least fourteen other mineral products, is no mean achievement. Still more impressive are Canadian percentages of world production: 80 per cent of nickel, 60 per cent of asbestos, 40 per cent of platinum, 20 per cent of aluminum, 18 per cent of zinc, 12 per cent of gold, and nearly 10 per cent of each of copper and lead. In the last fifteen years volume has increased about three-fold, and in the last ten, value has risen from $500 to $2,000 million dollars. In 1955 minerals or ores constituted 25 per cent of our commodity exports.

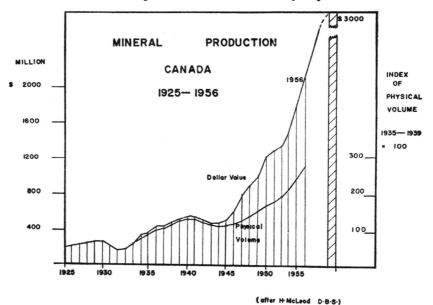

(after H McLeod D·B·S·)

FIGURE 2.

The changing complexion of the mineral industry is equally startling. Great as has been the expansion in the mining of long-famous nickel, copper, gold, and asbestos, and in production of aluminum from imported ores, this is being exceeded in rate and value by petroleum, natural gas, uranium and iron, admittedly in response to tremendous investments and both domestic and foreign demand. Within a year, uranium production[5] is expected to increase two-fold; in eight years, iron, three-fold; and within 25 years, petroleum and natural gas production, six-fold and seventeen-fold, respectively. With increasing population great growth may safely be predicted in production of essential

[5] A. H. Lang, "Our Uranium Resources," *Canadian Mining Journal*, LXXVII (1956).

industrial minerals—cement, sulphur, gypsum, salt, potash, and structural materials. What then of our reserves to satisfy such accelerating rates?

Our Known Mineral Reserves

In estimating reserves of mineral resources a very rigorous practice is followed so as to distinguish between what is really proven, what is probable, and what may be inferred from scanty data. Only the first two categories will be considered now, but it must be emphasized that these give a very incomplete picture of the over-all mineral possibilities, good as it may be. The following three tables, IV, V, and VI, present a minimum of statistics and show the expectable life of ten important resources, calculated from conservative estimates of tonnages or other units and the 1956 or planned rates of production.

Here we see that coal,[6] bituminous or of lower grade, at the present rate of mining will last some thousands of years. Depending on efficiency

TABLE IV

MINIMUM LIFE OF RESERVES AT PRESENT OR FUTURE RATES PRODUCTION: FUELS

	Reserves, 1956	Recovery	Rate	Years
Coal	49,000 million tons	50%	20 million tons (1950)	2,500
	70,000 million tons	70%	20 million tons (1950)	3,500
Petroleum 4.3 billion bbls.		+ 1/3 billion bbls. (annual increment)	170–1000 million bbls.	20–25+
Petroleum 1–100? billion bbls. (tar sands)				± 100
Gas 23 trillion cu. ft.		+ 2 trillion cu. ft. (annual increment)	0.17–3 trillion cu. ft.	173–60 (Alta. only)

of recovery this figure may be raised or lowered. Tremendous development of petroleum and natural gas, still going forward, demands that estimates of life be based not just on known reserves but on reliable predicted rates of discovery and rapidly mounting production over the next 25 years. A proven reserve of 4.3 billion barrels of petroleum,[7] it is reliably forecast, will be increased annually by an average of ⅓ billion barrels for some years to come, while production is likely to

---

[6] B. R. Mackay, *Coal Reserves of Canada: Report of the Royal Commission on Coal,* 1946; and articles on coal and energy by C. L. O'Brian and associates, *Transactions of the Canadian Institute of Mining and Metallurgy,* LIII, LV, LVI, LVIII, LIX (1950–6).

[7] "Oil Reserves in 1956," *Myers' Oilweek,* Vol. 8, (February 22, 1957), p. 49.

increase from 170 million barrels to 1,000 million. Similarly with gas, a reserve of 25 trillion cubic feet (? underestimate) is likely to increase by at least 2 trillion cubic feet annually. While production is measured now in modest billions of cubic feet, by 1982 it may be as much as 3 trillion. This cannot, of course, go on indefinitely. Ratio of reserves to production are under constant review and can be controlled, and, as we shall see later, the life of both oil and gas deposits is likely to be extended far into the future.

Iron and titanium ores are good for some hundred or more years and such metals as zinc, lead, copper and nickel for some forty to fifty

TABLE V

MINIMUM LIFE OF RESERVES: METALS

	Reserves (tons) 1956	Rate (tons or $)	Years
Iron ore	5,277 million + (× 2 or 3?)	30 million	150+
Titanium ore	150 million +	396,000	400
Zinc	23 million	423,600	54
Lead	9.9 million	186,674	53
Copper	14.5 million	353,292	41
Nickel	6.4 million (× 2?)	177,993	35 70
Uranium	$1,484 million +	$39–$400	7+

years. These, I must emphasize, are minimum figures which will only hold *if we find not another ton of ore*. The picture for uranium is somewhat similar. Reserves worth close to $1½ billion have rapidly been developed in the last five years. By 1959 production should jump in value to between $300 and $400 million, and although at this rate the life of present reserves will be short, as with other metalliferous deposits, the potential is great.

To this list should be added unlimited or really large supplies from which the light metals—aluminum, magnesium and lithium—may, if the need arises, be obtained. Included too are vast deposits of industrial minerals of a wide variety, such as asbestos, gypsum, salt, potash,

TABLE VI

MINIMUM LIFE OF RESERVES: INDUSTRIAL MINERALS

	Reserves (tons) 1956	Rate (tons)	Years
Asbestos	958 million	1–17 million	56+
Potash	5,000 million	?	5000+
Gypsum	1 billion +	4.8 million	200
Sulphur	Vast (gas, sulphides)	1.2 million	?

sulphur. Less important in quantities needed, but available in excess of demand, are such metals as cobalt, molybdenum and even mercury. Lacking in abundance or quality, but not in all cases entirely unknown, are deposits of many of the ferro-alloys—manganese, chromium, tungsten and vanadium; tin and some of the newer (atomic-age) metals such as beryllium, boron, zirconium and niobium. For these, we shall likely continue to depend on others for some time to come.

Our Potential Supplies

By this heading I refer to mineral resources still unfound but geologically likely to be discovered. Here we can only paint with a broad brush. In approximate order of probability, first will come additional reserves of metals in active mining camps as exploration proceeds to greater depths.[8] There is an old adage that the best place to prospect is in known mineralized areas. Here reserves may be increased by mining maximum amounts of marginal or sub-marginal ores along with higher grades, or alternatively, although not always possible, so planning that sub-marginal ores may be recovered later, when prices rise.

Then may be mentioned such known favourable hosts as the bituminous sands of Athabasca[9] with possibly hundreds of billions of barrels of oil and other by-products, innumerable iron-bearing formations, as in Labrador-Quebec and other areas in Ontario, the Northwest Territories, and even Alberta, deposits which lend themselves to beneficiation; the promising new nickel-copper belt across the northern end of Ungava; possibly copper of the Coppermine River country (N.W.T.); about two-thirds of our unmapped Precambrian Shield, favourable to so many of the metals; and finally, the incompletely explored areas of the northern Cordillera, the Appalachians of Quebec, New Brunswick and Newfoundland, and even parts of the Precambrian basement underlying Palaeozic and younger cover rocks.

The prospect truly seems unlimited, even if deceptively so, especially when we couple with it new methods of rapid mapping, new techniques and instruments for the discovery of ores, and faith and confidence in the efficiency, ingenuity and daring of our great mining and petroleum organizations and scientifically trained prospectors. What is there to be found, will be found. What is there to be mined and extracted will be forthcoming, as thoroughly and as efficiently as it is possible to do. All

[8] Maximum depths to which mining may extend will probably not be much over 12,000 feet. Wells drilled for oil, however, have attained about twice this depth.

[9] Department of Mines and Resources, *Drilling and Sampling of Bituminous Sands of Northern Alberta, 1, Results of Investigations 1942–1947* (1949).

this will take time. It will be costly. It will require much fundamental research not only by geologists, but by chemists and physicists who must be recruited in greater numbers by the industry. Such factors as these, as well as natural frontier barriers, will of themselves slow down the exhaustion of our resources, some of which will clearly long outlast others.

Problems and Policy

Great as the outlook may be it is not one for complacency nor a policy of *laissez-faire*. We clearly will have, for long years to come, far more of a wide variety of mineral resources than we can possibly use ourselves, resources which the rest of the free world needs. We have long been dependent on others for coal in central Canada, for oil and its refined products, for bauxite for aluminum smelters, and for most of the important ferro-alloys. We may decrease our imports, but we cannot cease, though we may control, our exports. No sound argument can be advanced for conservation of resources "simply for the sake of hoarding,"[10] but we can plan for their discovery and orderly development, we can prevent high-pressure exploitation, endeavour to maintain a suitable ratio of reserves to production, avoid waste and over-production, urge greater use of the more abundant and less of the scarcer materials (a world-wide problem), stock-pile strategic minerals, and assure adequate domestic refining facilities where justified, so that the maximum benefits from these will accrue to all Canadian citizens.

That there are problems can only be illustrated by a few examples. Foremost is the general problem of replacing exhaustible reserves by new discoveries. This may be left largely in the hands of private enterprise, supported increasingly by both reconnaissance and detailed mapping by the Geological Survey of Canada and provincial departments of mines, and by much-needed basic research on the occurrence of such resources.

More specific problems must also be faced. Disadvantageously placed, caught in a squeeze play between oil and gas, and soon, uranium, the coal industry reveals the striking paradox of depressed sections in the midst of a booming economy. Gold mining, always a stabilizing influence, is stymied by factors beyond our national control. Gas—for years wasted at the well-head, now being channeled into pipe-lines—affords an example of confused and too rapid planning and decision at top level, which will detract from the great boon it will be. Provincial control poses

10 Anthony Scott, *Natural Resources: The Economics of Conservation* (Toronto, 1955).

problems in natural resources (as well as in education), such as diverse regulations, staking and concession rights, and sectional as against national interests or control, as, for instance, the export of iron from British Columbia to Japan, preventing its conservation as a basis for a western iron and steel industry. Many other examples might be cited; all point up the need for long-term, careful planning.

How best may this be done? Not, certainly, by any one individual, nor yet by a multiplicity of agencies, federal and provincial; rather, would I suggest, it might best be undertaken by a national minerals policy (or resources) board—call it what you will. It should deal with *all* minerals, not just energy resources. It should be *continuing*, not periodic. It should be primarily advisory. Drawn from the chief sections of the industry, co-operating with the many efficient government departments now dealing with such matters, it should study, plan and co-ordinate the many phases and problems of our mineral resources, and above all it should be articulate and authoritative as to what is best for the nation as a whole.

"Take no thought for the morrow" is a good biblical precept. Many, not knowing what the future may hold as we pass into an atomic age, would probably follow it. It has not been adhered to in the present case, but we are thinking not of ourselves, but of generations to come.

Acknowledgements. Preparation of this paper would not have been possible without the very valuable assistance of others. Particularly thanks are given to Dr. J. F. Henderson, Dr. George S. Hume, Dr. John F. Walker, Dr. B. R. MacKay, Mr. W. Keith Buck, and Mr. C. L. O'Brian. Thanks are also due Dr. J. W. Ambrose for a critical reading of the manuscript.

L. M. PIDGEON

METALS FORM THE BASIS of an industrial society. Their extraction from ores and subsequent conversion into useful alloys is the field of the extractive and physical metallurgist. Ores are located by the prospector with the help of the geologist and geophysicist, brought to the surface by the miner, and turned over to the metallurgist for reduction to the metallic state. The ability of this team to locate ores and produce metals has increased enormously in recent years, but so has the demand until the tonnage of metals produced in the present century exceeds the sum of all that of preceding history. This colossal demand has, in many cases, exceeded the rate of discovery of new reserves, and concern is expressed that we are outstripping our resources. What is the responsibility of the metallurgist, and what is he doing about it?

It can be shown that a very substantial credit has been established. Methods of production of the classical metals have improved so that lower-grade deposits may be worked, but most important of all the metallurgist has learned to produce metals which were not available in the past. All the metals on the earth's crust may now be produced and, in fact, some new ones that did not exist. Even the so-called non-metallic minerals have yielded metallic silicon, titanium, magnesium and aluminium.

These reactive metals were denied to previous ages by inherent technical difficulties. The history of metallurgy begins with the metals, gold and silver, exceedingly rare, very stable (attacked only by the king of waters "Aqua regia"—nitrosyl chloride), and very heavy. Later, still in prehistoric times, copper, lead and finally iron, were produced. The latest stage of metallurgical history is the production of light, reactive metals, such as aluminium and magnesium, relatively unstable, but present in abundance. Thus, the first metals were the most rare, and many of the latest are the very stuff of which the world is made. This paradox may be usefully examined.

TABLE I

Metal	Date of first "commercial" production	Estimated abundance in earth's crust	Free energy of formation of oxide per atom of oxygen @1000°K
			kilogram calories
Magnesium	1905	2.1%	−117.7
Aluminum	1880	8.8	−108.2
Uranium	1944		−109.0
Titanium	1946	.3	91.1
Chromium	1900	.02	69.7
Zinc	1600	.004	− 58.9
Iron	*c.* 1500 B.C.	5.1	− 47.8
Nickel	1870	.01	− 35.0
Tin	*c.* 1500 B.C.	.004	− 43.6
Lead	*c.* 2000 B.C.	.001	− 28.7
Copper	*c.* 3000 B.C.	.01	− 23.3
Silver	Prehistoric	.00001	+ 7.7
Gold	Prehistoric	.0000005	+ 10.0 (500°K)
Carbon			− 47.9

In Table I some of the best-known metals are listed, together with the approximate date of their commercial production. Additional columns show their estimated abundance in the earth's crust and the free energy of formation of their oxides at a given temperature. This latter value is a measure of the chemical work which must be performed to separate the metal from its oxide. It is at once apparent that, until modern times, no metal was produced with an oxide exhibiting a free energy of formation greater than —47.8 kilogram calories (at 1,000°K or 1,340°F). The ancients could not surmount the energy barrier represented by higher values.

According to definition, a compound which is formed during a negative free energy change is stable, and the greater the negative value the greater the stability of the compound and, of course, the greater the difficulty in breaking it down into its component elements. Conversely, a positive value represents instability and a tendency to break down. At 1,000°K, therefore, magnesium oxide is the most stable oxide shown, while gold oxide is unlikely to exist. Even silver oxide, stable at lower temperatures, is at this temperature broken down into silver and oxygen.

Thus, gold and silver exist as metals and not as compounds at 1,000°, and even at ordinary temperatures this is the case. Thus, these metals required no special reaction to produce the metallic state, and they were the first metals to be used by man.

Their occurrence is exceedingly rare, however; hence, the prospector has ever been and still is more important than the metallurgist, where these metals are concerned. Modern improved metallurgy has permitted the recovery of gold from very dilute ores, but mining costs are still a limiting factor.

The oxides of metals above silver (Table I) are stable, even at elevated temperatures, and a new device was needed to separate the metal. The discovery that carbon would remove the oxygen took place at the beginning of recorded history. The basic smelting reaction

$$MO + C = CO + M$$

was effective in producing metals up to iron (Table I) by 1500 B.C. For metals listed above iron in the table, however, the reaction is much more difficult to apply. Zinc is a comparatively modern metal, while many metals at the head of the list cannot be produced at all by this method.

The simple carbon reduction, like all reactions, must exhibit a negative free energy change. Reference to the last column of Table I shows that this is just possible at 1,000°K with metals up to iron ($\Delta F°_{1000}$ FeO = —47.8, CO = —47.9). The reduction of iron oxide by carbon represented the limit of ancient techniques which relied on combustion of part of the carbon to produce this energy.

Metals above iron can be reduced by carbon, but at much higher temperatures ($\Delta F°$ for CO increases its negative value with temperature, while the value for all metallic oxides decreases). Zinc requires a minimum temperature of 1130° for reduction, and this temperature is near its boiling point. The difficulties of handling gaseous zinc were not solved in Europe until modern times. Metals above zinc require temperatures and techniques which have only been available quite recently. Thus it is that the ancient Egyptian produced all the common non-ferrous metals, including iron, even zinc is "early" modern, and only in recent times have all other metals been produced. Their production has followed the discovery and development of electrical energy which has freed the metallurgist from the limitations of temperature and atmosphere imposed by the use of carbonaceous fuel firing.

Thus, the earliest metals to be produced were also the most rare. Only iron was present in large quantities. The most plentiful metal, aluminium, could not be separated from its ores and, until recently, such ores were called "non-metallic minerals." Our first debt to the future we have already paid. We have been able to produce metals from all the compounds on the surface of the earth. There are now no "non-metallic" minerals. This achievement offers us metals which exist in inexhaustible abundance. Aluminium, in the form of clay, at present an uneconomic ore, will not be depleted by any effort of man. One cubic mile of sea-water contains enough magnesium to supply the present world demand for some twenty years, after which the next cubic mile may be tackled. Titanium is also a plentiful metal, although ores amenable to present extraction techniques are less widespread.

While some of the reactive metals near the top of the table are very plentiful, they all form very stable compounds. While the modern techniques can break down these compounds and liberate the metal, they do so only by the expenditure of large amounts of energy. The free energy of formation is an indication of the relative value of this work. The actual work demand is even greater than that suggested, because the work must be supplied at a high level under inherently wasteful conditions. Aluminium requires the expenditure of 10 kilowatt hours per pound, while copper may be produced for less than one kilowatt hour per pound.

In this case, therefore, energy is the consumable and not the ore reserve. In Canada, we already possess a large aluminium industry which operates upon imported ore, and exports most of the metal produced. The industry is located in Canada solely because of the energy. From this point, the discussion could proceed to a consideration of world reserves of energy, but this would be outside our terms of reference. It may be suggested, however, that, as all available hydroelectric sites are utilized, the production of energy will fall back on fossil fuels and eventually nuclear energy. If this process tends to level the cost of power throughout the world, it seems likely that aluminium and magnesium production, since ores are ubiquitous, will take place close to the points of greatest consumption.

If we have opened the way to great energy resources, we have also offered unlimited supplies of certain metals at least.

What is our debt to the future in the classical metals: iron, lead, zinc, copper, including nickel? The demand for these metals has increased enormously, and concern is frequently expressed as to future supplies. With the exception of iron, these metals are rarities, and depletion of workable deposits is inevitable if a sufficiently long time scale is chosen. A number of factors are in sight now which will put off this day: (*a*) improved techniques for the location of ores; already magnetometric and seismic methods have widened the reserves, and this may be expected to continue; (*b*) improvements in extraction techniques enabled low-grade ores to be profitably explored; (*c*) increase in cost of essential metals; (*b*) and (*c*) are of course related; (*d*) more effective recirculation of scrap; (*e*) development by the physical metallurgist of alternative alloys which are effective substitutes for rare metals. These possibilities exist at present and their importance will be enhanced by ultimate scarcity.

Copper and nickel appear to be irreplaceable at present; the former as an electrical conductor in motors and the like, and the latter as an essential ingredient of acid- and heat-resistant metals. Even here, however, competitors are in sight. Aluminium has already captured the heavy conductor field, while zirconium and titanium exhibit excellent corrosion resistance, but are very costly.

The great demand for iron has already depleted many high grade deposits. Entirely successful methods, however, have been developed to beneficiate lower-grade deposits, which are abundant. Iron, ever the most important metal, is unlikely to be depleted. We owe the future nothing here.

In conclusion, the past half century has seen a great increase in the use of metals. In some cases, the demand shows signs of outstripping the development of new deposits. On the other hand, the same years have seen the discovery of methods of producing all metals. Some of these are the most plentiful in the earth's crust and cannot be depleted. One of these, aluminium, is already second to iron in tonnage. The only penalty is the heavy demand for energy which is required to separate these metals from their ores. Energy is in fact the only ultimate consumable. The new metals must compete with other demands for energy. Given adequate supplies of energy, iron, aluminium, magnesium, and perhaps titanium are assured for all times. The vital non-ferrous metals, copper and nickel in particular, are ultimately to be in short supply, although 1957 prices do not suggest this. In fact, it is estimated that consumption of nickel must be increased by 75 per cent to accept the present production capacity. Rare and special metals will take their place in the race of supply and demand. The user who pays the price must take his choice. The physical metallurgist will no doubt supply alternative metals when demand for one far exceeds supply. The future of metals seems secure. The problems are surely less severe than many others facing the world, such as food, water, and radiation hazards.

B. S. KEIRSTEAD

WE HAVE BEEN CONSIDERING, in this symposium, "our debt to the future" and I want to begin my discussion now by thinking of our debt to the past. True we are creating new capital in Canada at a fantastic rate, something like 26.1 per cent of Gross National Product, but one of the reasons we are able to do so is because we are well equipped with productive capital which was laid down by past generations. I shall not take time to enumerate in any detail our inheritance of capital, "our debt to the past." I shall mention only our transport system. By 1890, or at the latest 1913, the major capital outlay on the right-of-way was complete for our national railway system. Canadian economic development of this century, indeed Canadian national existence, would have been impossible without this great capital creation of our forebears. We are creating a new and extended St. Lawrence Waterway. But we build upon a waterway, dredged up to Montreal and extended west by a massive system of canals, which was largely complete by the turn of the century.

In turn, we, of our age, are creating capital which, added to what we already possess, will be our bequest to the future. What is "our debt" in this respect? How much capital should we be creating for our children and our children's children? To this question, which is both economic and, in a sense, moral, I ask that we now turn our attention.

I think the first point to be made is that, though I have perhaps presented the question in an academic way, it is anything but an academic question. Certain people, governments and business enterprisers, have the responsibility of deciding how much new capital to create. Quite different people, you and I and consumers generally, decide how much, taken together, we want to consume and, conversely, to save. All our existing productive resources—labour, natural resources and present capital—must be used, to the extent of full employment, either in making goods for consumers or in making additions to our stock of capital. Thus there is a kind of competition between capital goods production and consumers' goods production. We cannot have as much of both as we might like. A rate of economic progress which might satisfy our pride as citizens, the ambitions of our politicians and the more optimistic expectations of our business enterprisers, may well be a rate which leaves too little for the present production of consumers' goods. By "too little" I mean simply "less than consumers are trying to buy."

This competition—between capital goods and consumers' goods—is exactly what has been taking place in Canada in the recent past and the present. The indication of it is price inflation. Inflation is a manner of forced saving. It is a way of preventing people from consuming, because of high prices, so that government and business enterprise can create the new capital—in the form of highways, CF-105's, new mines, factories and shopping centres—that they want.

Thus we come to the heart of the problem, the question posed for us by the general title of this symposium, "our debt to the future": what should we lay down in the way of new capital? Obviously, from the degree of inflation present in Canada, there is a difference of opinion about this. Canadian consumers do not want to make the capital provision that Canadian governments and Canadian businesses are trying to make. This consumer reluctance is evidenced by price inflation. You and I—and a lot like us—just do not want to save enough, that is to go without present consumption, to provide for the capital creation which our civil servants and business executives have decided upon. Our resistance and their stubborness give us this competitive bidding for the services of all productive resources, which we call inflation.

Let us now turn to the answer economists have worked out to our question, viz., how much should a society save? (Or, if you prefer, how much new capital should a society attempt to create?) We cannot claim that the economists' solution is satisfactory, or that it is fully agreed. Nor do I suppose that I can do justice in about two minutes to the elegance of Irving Fisher's mathematics.

The received economic answer to our problem—how much new capital should be created by a society at any time (or what is our debt to the future?)—is that this should be determined by the market mechanics. Let me explain a little further. It is a matter of comparing future goods with present goods. According to the economists I am

citing—or, rather, oversimplifying—there is some rate of time prefer-
ence, or some rate at which future goods are discounted as compared
with present.goods. Thus an increased supply of consumers' goods in
the future must be discounted when compared with an immediate supply
of consumers' goods today. Capital must be regarded simply as a means
to an increased supply of future goods. It increases productivity. In
the present, of course, it means, as we have seen, a forgoing of
consumption. It is in this sense that the "market mechanism" comes to
operate. From the point of view of the business man, who makes the
investment in new capital, the increase, or expected increase, in productivity must pay for the interest he has to pay on the borrowed money
with which he buys the new productive capital. And you will remember
that by "capital" I mean real capital, some physical assets. The interest
rate is supposed to measure the average rate at which people in the
society discount future goods. So there it is: there is a rate at which
the future is discounted, this rate determines the rate of interest on borrowed funds, or the rate at which people are willing to forgo present
consumption in the hope and expectation of increased consumption in
the future, and, as long as new capital is expected to provide this
increase in consumption by its increased productivity, new capital will
be created. When the rate of increased productivity of new capital falls
below the rate at which future goods are discounted, it is obviously uneconomic to create new capital.

Thus, in summary, economists traditionally depend on the rate of
interest, as established in a free market, to determine the amount of new
capital to be laid down by any generation.

I am forced to be almost completely negative as I assess this traditional argument. To begin with, I do not myself really accept the
"received" economic position. It claims to be ethically neutral, when it
is not. Its claims to ethical neutrality rests on the view that, whatever
the rate at which the present generation discounts the future, that is
the "right" rate. Thus, for six centuries, when no capital except churches
was created, this zero rate of capital accumulation in materially productive activity was the "correct" rate. Then suddenly capital began
to be created in England at a rate of more than 10 per cent of Gross
National Product per annum. This also was a "right" rate. Clearly, the
basic philosophy here is that with which Voltaire had so much fun in
Candide: whatever is, is right, or this is the best of all possible worlds.
The fundamental ethical assumption of nineteenth-century capital theory
was, in fact, that the sum of the decisions of the mass of consumers and
savers was the only possible objective criterion of what should be saved.

Clearly some other criterion, historical or ethical, becomes necessary.

In the second place, I doubt if either the rate of interest or the rate of increase in the productivity of capital (what is called the "marginal productivity" of capital) has much to do with the rate of new capital creation. I am not going to elaborate this point here. The empirical evidence is that the rate of interest is determined by acts of government policy and that new investment is undertaken in the expectation of profit far in excess of either the rate of interest or anything which could be called the "marginal productivity" of capital.

Yet, when we examine alternatives to our market mechanism, however unfree it has become, we observe that there are grave dangers connected with any "planning" for the future. The market mechanism, whatever its defects, at least had the merit of objectivity. The division of resources between the production of goods for present consumption and for future use was a division which was not imposed by some one person or party. If we plan for provision for the future, presumably some group, or even individual, who makes the plan must impose his judgment of what is a right or proper decision upon the rest of us. I suppose in a parliamentary country there would be some popular control over these decisions, but I am not sure how great it would be. We already have examples of "selling" the people a very rapid rate of economic development, which means a high rate of provision of new capital. Consumers' efforts to protect themselves against these decisions end only in inflation. Nor are votes likely to be effective. Political parties, when seeking power, promise to spend more rather than less.

There is very little danger that we in Canada will make too little provision to meet our debt to the future. Private enterprise and governments are making vast provision today. I suspect that in addition to private capital responding to market forces, we shall see some planned capital creation as governments try to encourage economic development in regions which have lagged behind the richer parts of Canada. Though such planned development will doubtless, in our Canadian tradition, be *ad hoc* rather than "over-all" in nature, we can scarcely believe either that some regions will be left to rot while others progress, or that present consumption demands will significantly retard the rate of economic progress.

We have, so far, said nothing about the connection between the creation of new capital and technical innovation. These two processes are so closely related in fact that Adam Smith failed to distinguish between them. In a sense, I think he was right. More careful economic logicians have made of the distinction almost a separation. In fact, the two processes are never separate. With our resources, and with our own advancing technology, to say nothing of what we derive from the United States and the United Kingdom, we are committed, whether we like it or not, to a period of rapid capital creation and renovation. Consumer reluctance cannot prevent this development. All it can do is influence the rate of change. In this regard, I expect we shall work out a typical Canadian compromise. I suspect that this kind of rough and ready compromise between the imperfections of market-directed capital creation and fully "planned" development is less to be feared than either alternative.

Finally, I believe that fiscal and monetary policies designed to check inflation are necessary to protect the consumer and his so-called sovereignty. When these policies are severe enough to be effective, they hurt certain classes and they prevent certain businesses from carrying out their intended investment programme. That is exactly what they should do, and when they hurt some considerable section of the business community, government should stand to its guns and not back down in the face of business pressure. With firmness and common sense, government can see that our debt to the future is fully met by a sound rate of development, but it can prevent that rate becoming too great for the masses of Canadian consumers, which is to say the masses of Canadians alive today. It is good to remember what we owe to the past and to decide to pay that debt by ensuring a good life for Canadians of the future. It is also good to see to it that all our citizens living today have an opportunity to live good lives.

MAURICE LAMONTAGNE

ECONOMISTS have never been especially interested in the problem of the conservation of resources. They have taken natural resources more or less for granted or they have regarded them as capital goods which did not deserve any special consideration. Economic textbooks very seldom mention the problem of utilization or conservation of resources. As a result of this indifference, the scientists connected with land, forestry, fisheries, water power, petroleum and minerals, and primarily interested in the technical aspect of conservation, have been inclined to extend the scope of their own inquiry into the realm of economics, often despite the lack of adequate analytical tools. They have developed their own economic theory of conservation and their own views on the kind of economic policies which should be implemented in this field.

Thus we are faced with two different and often conflicting economic theories of conservation. One is presented by the resource specialist as a special theory. The other is offered by the economist but it is only a special application of the general theory of capital and investment, and it has not been systematically developed. Since the main purpose of this paper is to initiate a discussion between the resource specialist and the economist on conservation, it will be limited to a brief presentation of their two different lines of approach to this problem.

The approach usually followed by the specialist can best be described in general terms by the following definition of conservation presented by R. T. Ely: "Conservation, narrowly and strictly considered, means the preservation in unimpaired efficiency of the resources of the earth or in a condition so nearly unimpaired as the nature of the case, or wise exhaustion, admits."[1] According to Professor Scott Gordon, biologists in their approach to the fisheries problem focus their attention "on the quantity of fish caught, taking as the human objective of commercial fishing the derivation of the largest sustainable catch."[2] Indeed, this concept of a high sustained yield seems to be commonly accepted as a desirable economic end by all resource specialists, especially in the field of renewable resources. This approach, therefore, is defined in purely physical quantities, not in economic terms, but it has several important economic implications. Its basic principle is that the sustained yield (expressed in physical terms) to be derived from natural resources should be maximized and this principle is presented as the main economic objective.

When the concept of a high sustained yield becomes the main objective of economic activity and policy, natural resources are considered as the only scarce factor of production; it is thought that the industrial structure of a region or a nation should be planned and adjusted not only to enable the resources to attain such a high yield but also to make full use of it. The other factors of production, namely labour and capital, are more or less taken for granted and they should be so utilized as to facilitate the achievement of the major goal. A rate of utilization lower than the maximum level on a sustained basis is considered as a waste, even if that maximum level involves insufficient returns for the other factors of production. For instance, arable soil should not be left uncultivated even if farm income is not attractive, and even if government subsidies become necessary to supplement such income. On the other hand, a reduction in the stock of resources is undesirable *per se.*

---

[1] R. T. Ely, ed., *The Foundations of National Prosperity* (New York, 1918), p. 13.
[2] H. Scott Gordon, "The Economic Theory of a Common-Property Resource," *Journal of Political Economy*, XLIV (April 1954), p. 128.

Indeed, if the existing stock is being depleted, restrictive regulations should be introduced, current production and employment should be curtailed and more capital should be invested to increase the stock. If private capital is not available for that purpose, the government must intervene and undertake a conservation programme. Imports of raw materials may be not only desirable but essential as a conservation measure, especially in the field of non-renewable resources. On the other hand, natural resources should not be exported but fully processed or utilized in their country of origin.

This brief description may be oversimplified but it represents in substance the typical approach of the resource specialist to the economic problem of conservation. Professor Scott Gordon has shown some of the limitations of such an approach to the fisheries problem. His criticism, which is of general application, is stated as follows:

> Focussing attention on the maximization of the catch neglects entirely the inputs of other factors of production which are used up in fishing and must be accounted for as costs. In fact, the very conception of a net economic yield has scarcely made any appearance at all [in the biological literature]. On the whole biologists tend to treat the fisherman as an exogenous element in their analytical model, and the behaviour of fishermen is not made into an integrated element of a general and systematic bionomic theory.[3]

It has been said that this approach seeks "to economize on human welfare in order to achieve the maximum welfare for resources." It would perhaps be truer to say that it is based on the assumption that natural resources are "unique and irreplaceable" and that investment in their conservation is an absolute necessity. This conception of economic life is highly static. It does not visualize that technological innovations could reduce the usefulness of a given resource or provide a substitute for it, and that labour and capital applied to the sector of resource industries could have better alternative uses.

The approach to conservation of resources followed by economists has been in the opposite direction. The late Professor Innis described this approach in a rather extreme form:

> The drive of modern technology with the modern pecuniary economy involves exhaustion of natural resources and getting on to something else. Depletion of pulpwood enables hydro-electric power to be turned from paper plants to other industries in the interests of "progress" and "higher standards of living". The problems of conservation are concerned with restricting technology as well as with improving it and utilizing it to capacity.[4]

3*Ibid.*, p. 128.
4H. A. Innis, "The Economics of Conservation," *Geographic Review*, XXVIII (1938), p. 137.

Classical economists considered most of the natural resources under the broad term of "land" as being an exogenous factor in their analytical system. According to Ricardo, land consisted of "the original and indestructible powers of the soil"; and Marshall added, "man has no control over them; they are wholly unaffected by demand; they have no cost of production, there is no supply price at which they can be produced."[5]

Modern economic theory has abandoned the distinction between "land" and "capital" because it has been found that few of these "permanent and indestructible" gifts of nature were in fact inexhaustible. Most of the natural resources can be destroyed and replaced by man. Since they can be depleted by use and increased by giving up current output for future output, they can be treated as capital goods and their conservation as a type of investment.

Professor Anthony Scott expresses this view as follows:

The exploitation of natural resources is merely a special case of the using up of any productive asset; and the prolongation of their use is governed by the same general principles as govern the depreciation and maintenance of machine and buildings. . . . Conservation of resources is not only analytically analogous to investment in capital goods; it is also, in each planning period, an actual alternative to investment in and consumption of available goods and services. Society must constantly choose how it will allocate resources among competing uses and among periods of future time. Conservation of resources is only one of many possible choices; one that implies sacrificing present consumption and investment in produced means of production in favour of increased future supplies of a group of natural resources. Given an aggregate amount of savings available for investment in each period of time, increased conservation of resources must mean also a reduced endowment of buildings and equipment for posterity. It is ridiculous, then, to say that conservation is a movement which has the welfare of the future particularly in mind; conservation will not necessarily increase the future's inheritance, but merely change its composition from "capital goods" to natural products.[6]

Thus, the factors governing the capital-formation process in a dynamic economy also determine the degree of maintenance of natural resources. In a price-directed economy, in any given period of time, investment in conservation will be determined by the price of borrowing money, which includes, on the one hand, the interest rate, a risk premium and the amortization of underwriting costs, and, on the other, the net economic yield expected from that investment compared with net prospective returns on alternative types of investment. Economists contend that if,

---

[5]Alfred Marshall, *Principles of Economics* (8th ed., London, 1920).
[6]Anthony Scott, *Natural Resources: The Economics of Conservation* (Toronto, 1955), p. viii.

under a free enterprise system, all the factors governing this market process were allowed to work in real life as they behave in their theoretical models, private businesses would adopt only those conservation measures which really maximize human welfare, and that the problem of conservation would be adequately solved by society through the price mechanism without government intervention.

In their theoretical models, economists have defined the equilibrium condition of the economic system so as to maximize net economic or monetary yields. They assume free competition, perfect knowledge and the absence of frictions—in other words, that the economic system is fluid and that labour and capital can always be easily shifted from one kind of employment to another. They also assume that the stock of natural resources existing in the world is quite unlimited, that technological progress tends to make these resources more and more interchangeable, and that free trade makes them readily available to all nations.

Thus, we are faced with two conflicting views on the economics of conservation. The resource specialist aims at maximizing the sustained physical yield to be derived from natural resources; he does not rely on the price system to achieve this objective and advocates a conservation programme sponsored by the government. The economist seeks to maximize monetary yields and contends that the proper functioning of the price system would maintain the utilization of all capital goods, including natural resources, at the socially optimum level. However, he knows that the conditions of the real world do not always ensure the proper functioning of the price mechanism and he views the role of government in this field as designed mainly to eliminate the disrupting influences which appear on the market.

Such conflicting views inevitably lead to conflicting policy recommendations. In this field, there seems to be agreement only on the necessity for society to educate owners and users of natural resources. The perfect knowledge assumed by economists seldom exists in the real world but ignorance leads to wasteful decisions and irrational actions which disrupt the proper functioning of the price mechanism. Research and education campaigns undertaken by the government contribute to improve this situation.

In other sectors of policy, however, there is a wide range of disagreement. Such disagreement is perhaps best illustrated by the proposals made in respect to common-property resources such as commercial fishing, trapping and hunting. Here the situation is characterized by the fact that the resources belong to everybody and the old dictum says that everybody's property is nobody's property. For the individual firm in these industries, the resources are free goods and, since it has no control over the market price of its products, it will normally try to maximize its physical production. Moreover, it cannot be expected to adopt conservation measures in order to prevent the depletion of the stock even if, under other circumstances, such measures could be viewed as a sound investment.

Resource specialists are inclined to blame the price mechanism and the profit motive for the conservation problem thus raised and they propose the adoption of restrictive regulations by the government in order to prevent depletion. They usually recommend legal limits to the catch or seasonal closure. Commenting on the application of this last method of control to fisheries, a biologist, Mr. W. E. Ricker, has this to say:

> This method of regulations does not necessarily make for more profitable fishing and certainly puts no effective brake on waste of effort, since an unlimited number of boats is free to join the fleet and compete during the short period that fishing is open. However, the stock is protected, and yield approximates to a maximum if quotas are wisely set; as biologists, perhaps we are not required to think any further. Some claim that any mixing into the economics of the matter might prejudice the desirable biological conse-quences of regulation by quotas.[7]

Professor Scott Gordon does not accept this conclusion and claims that "since the regulatory policies are made by man, surely it is neces-sary they be evaluated in terms of human, not piscatorial, objectives." Economists are inclined to think that these restrictive regulations are undesirable in terms of human welfare, because they have very little favourable effect on the prices of the product, but they lead to increased costs and reduced production and, consequently, to smaller incomes. Economists recognize that there is a problem of resource utilization and conservation in the sector of common-property resources, but they explain it by disrupting influences preventing the proper functioning of the price mechanism. An important disrupting factor is the fact that common-property resources are free goods for the individual producer and are valueless to him as long as he has not appropriated them; an individual firm will not try to economize them and, for all practical purposes, they lie outside the price system. When open resources are turned into property rights, the behaviour of the individual producer changes completely because they now have a value and a price for him. If they yield what he feels is an adequate income, he will try to

[7]Quoted by Scott Gordon, p. 133.

economize them and he will take steps to conserve them. From that point of view, it would be interesting to compare the behaviour and the income of the trapper and those of the rancher.

Another disrupting factor in these resource industries is that they are more easily accessible to the individual producer than other industries and that the individual is not inclined to leave them in spite of the low incomes they provide. Scott Gordon describes the great immobility of fishermen as follows: "Living often in isolated communities, with little knowledge of conditions or opportunities elsewhere; educationally and often romantically tied to the sea; and lacking the savings necessary to provide a 'stake', the fisherman is one of the least mobile of occupational groups."[8] The same observation could be made about trappers and hunters but especially about the Indians and Eskimos in northern Canada. This immobility, which prevents the proper functioning of the price system, leads to uneconomic levels of production, which may also be too high from the point of view of a conservationist policy, and to lower incomes than in other occupations.

Thus, the policy advice given by economists on the problem of utilizing common-property resources is not a conservationist programme which might prevent depletion but which might at the same time increase costs and reduce incomes. Their recommendation would be to turn open resources into property rights—for instance, to encourage herding rather than hunting, ranching rather than trapping—to restrict the entry into those industries and to encourage people to leave them. They claim that if policies of this type were carried out incomes would be raised and, at the same time, the conservation problem, if it existed, would be solved.

In the field of soil conservation, we are faced with similar conflicting views. The resource specialist, in his attempt to maximize the physical yield of the soil, is inclined to favour the extension of the area of arable land and to advocate reclamation projects. On the other hand, the economist is impressed by the current unsatisfactory situation of the market for agricultural products. He views most of these reclamation projects as undesirable types of public expenditure which, under present market conditions, lead to greater agricultural production and increased demand for subsidies designed to stabilize farm income.

These few illustrations are sufficient to show that the views of the specialist and of the economist on conservation very seldom coincide. This in itself is a serious problem which increases the difficulty of developing adequate policies in the field of resource utilization. The

[8]*Ibid.*, p. 132.

economist is probably largely responsible for this unsatisfactory situation. He has developed useful tools of analysis and he has applied them to the study of numerous economic problems but he has not yet used them to build a systematic theory of resources which could readily lead to appropriate policies. In the meantime the resource specialist has tried to fill this gap, but, since he is primarily interested in conservation by profession, his policy recommendations are probably better adapted to a stationary state than to the conditions of a dynamic economy.

HUMAN VALUES AND THE EVOLUTION
OF SOCIETY

G.-H. LEVESQUE, O.P., M.S.R.C., *Chairman*
THOMAS W. M. CAMERON, F.R.S.C.—A. S. P. WOODHOUSE, F.R.S.C.
R. ELIE, M.S.R.C.—ROY DANIELLS, F.R.S.C.

G.-H. LEVESQUE

DEPUIS TOUJOURS, pour les sociétés comme pour les individus, vivre, c'est évoluer. Mais jamais, dans toute l'histoire sociale des hommes, l'évolution n'a été plus marquée que depuis la dernière guerre. Jamais elle n'a connu un rythme plus rapide. Jamais elle ne s'est étendue en même temps à autant d'aspects de l'activité humaine. Jamais elle n'a manifesté son intensité chez autant de peuples à la fois.

Notre pays lui-même en est un exemple frappant, emporté qu'il est par le dynamisme de sa jeunesse, la puissance de ses forces économiques, la vertu de ses progrès techniques, l'élan de son développement culturel et scientifique, l'essor de ses institutions sociales et politiques, par ses obligations internationales enfin. Emporté, mais vers quoi ? Où cette évolution le conduira-t-elle ? Où devrait-elle le conduire ?

Car, l'évolution ne vaut pas par elle-même. Elle ne vaut que si elle conduit au meilleur, que si elle implique progrès. Continuité aussi : les deux pôles essentiels de toute évolution étant la continuité et le progrès. Evoluer, en effet, c'est rester mystérieusement soi-même tout en devenant autre; c'est marier le présent au passé pour qu l'avenir en jaillisse comme un enfant légitime et plus parfait. Entre tradition et progrès, il ne doit pas y avoir hostilité mais apparentement.

C'est pourquoi nous ne pouvons évoquer le problème de l'évolution sociale sans poser celui des valeurs humaines. Qu'advient-il à ces valeurs au milieu du flot évolutif ? Mais auparavant, que sont-elles ? Le produit ou la cause de l'évolution sociale ? Transcendent-elles toute évolution ? Y en a-t-il de permanentes qu'il faut sauver à tout prix et d'autres qu'on peut laisser tomber ?

Voilà autant de questions auxquelles l'avenir du Canada nous demande de répondre actuellement. Aucun expert, aucun penseur ne saurait prétendre trouver seul une réponse complète. Celle-ci ne saurait être que l'œuvre de tous, chacun apportant humblement mais généreusement sa part de lumière.

Cet après-midi, dans un modeste effort qui devrait être poursuivi sans cesse, nous demanderons successivement à un savant, le Dr T. W. M. Cameron, à un humaniste, le Dr A. S. P. Woodhouse, à un littérateur, M. R. Elie, et à un philosophe, le Dr R. Daniells, de projeter sur le sujet que je viens d'esquisser l'éclairage de leurs disciplines respectives.

As my final word, I wish to put an emphasis on Dr Daniells' striking conclusion: "The debate should go on" between humanists and scientists, the scientists trying to be as humanistic as possible and the humanists trying to impress upon themselves the discoveries of science, both working in close co-operation with a spirit of humility and with ambition as well. Humility will allow them to acknowledge the bounds of their respective disciplines. Ambition will entice them to work together for man and for all men, in order rightly to prepare the future of Canada.

THOMAS W. M. CAMERON

THE PUBLICATION of Darwin's *Origin of Species by Means of Natural Selection* around a hundred years ago had a far-reaching effect on human thinking. His theory has by no means explained fully all the observed facts and it has been subjected to much criticism and to even more misrepresentation. But there is no question concerning the establishment of its basic underlying truth, that all life is related. Man is no exception. Anatomically and physiologically he is a rather primitive type of mammal with a comparatively recent evolutionary history. He is distinguished from all other mammals by the great development of the frontal lobes of his brain, by the extent of his intelligence, and by his ability to create and to think both deductively and inductively, and to communicate with his fellows.

He is biological rather than logical in his actions, being fundamentally ruled by his instincts. He is much less rational than he thinks he is, and thinking is at best an intermittent process, most frequently employed when routine modes of action encounter obstacles. As with all other animals, pain and pleasure are the two ruling motives underlying his behaviour, and to most human beings material and non-intellectual pursuits will always predominate.

He is basically an unsocial creature, living in small family groups which, because of his innate helplessness and need for assistance, he of necessity has had to enlarge to form tribal units. Although he often lives in very large groups and has become civilized, he still is not really a social animal; it is problematical if he ever can be. Because he lives in company with his fellows, increasing density of population makes each individual more important to his neighbours, not only socially but physically. As Dr. J. B. Collip once expressed it: "The behaviour of the individual man is in part the expression of his conscious, his self-conscious and his unconscious life . . . and is determined by his hereditary background, his external environment and his internal environment."

Because he is not basically social, there is a continual conflict between his instincts and his reason, and successful standards of conduct set up by the society aim at the control—usually only partial—of his natural emotions. We may be able to make a man good by law but we cannot make him want to be good. Consequently, these extrinsic controls must always come into contact, and often conflict, with intrinsic controls exercised by his internal environment.

Social science is essentially a branch of biological science and, like civilization and culture, must rest on a clear understanding of man's biology. Standards of conduct, whether set by the society or the government which controls the society, must therefore be based on this and not on arbitrarily selected ideas which ignore his background. There can be no successful government by catchword.

The primary aim of social science should be to convert man into what the late Gilbert Murray once called "a decent Beast of Prey." It cannot confine itself entirely to his external environment. It must take into account his internal environment, which is under only partial mental control, the degree of which varies with individuals and which is to a considerable extent influenced by his heredity. Social science must realize more fully that psychology can never completely control physiology.

Man's internal environment is conditioned by both the central and the sympathetic nervous systems and the complex series of endocrine glands, the greatly varied functions of which are even yet very poorly understood but which play such an important part in regulating and maintaining not only the normal physical and mental reactions of the body, but the emotional as well. Very few people, even now, realize the essential part which the hormone (or endocrine) systems have in controlling the mechanism of the human machine. Probably the same essential hormones act in all kinds of vertebrates but in the human machine there is an important difference from all other animal machines. This is the enormous development of the frontal lobes of the brain which add self-consciousness to consciousness and, by enabling man to think at will, to modify—insofar as they can be modified—the results and even the actions of these systems.

Civilized man has to rely on an artificially fashioned external environment which requires both knowledge and techniques to maintain itself. His society is an artificial one which must be organized so as to overcome his solitary instincts and to minimize to others the effects of his individuality. It differs from all other vertebrate societies in that mankind has the responsibility of introducing into this society fully fledged citizens. This duty is implicit in the biological fact that each generation has passed to each succeeding generation an actual physical portion of the preceding one and, with it, the genes which are responsible for the transmission of the characters, both physical and mental, of both parents. However, the genes unite in the offspring only some of the characters of both parents who, in turn, have received some from each of their parents, and so on *ad infinitum*. Throughout all the previous generations of human beings, these genes have become so thoroughly mixed that it is statistically almost certain that in any given population of reasonable size, there will be similar proportions of individuals with similar potential abilities. We so readily accept the dictum that all men are equal that we forget that it is true only in the forensic sense and we fail to recognize how demonstrably false it is biologically. Quite obviously no two men are equally able but most can be competent in some particular fashion or at least possess the potential ability to be so.

It follows, therefore, that to have a balanced society the size of the population is important; equally important are the methods we use to bring out all the varying potentialities of this population. All kinds will almost certainly be required, but most urgently will be required those who are able to think constructively and to advance the culture of the society. This is the function of our educational system and the responsibility partly of the society, partly of the parents.

It is important to distinguish between education and training. The function of education is to make a life, that of training, to make a living. We have put so much emphasis on the latter that we tend to forget the former. Yet fundamental ideas come from the thinker—the educated man—while the practical man merely adapts them. Education begins in the home and is aided and continued in the schools but it reaches its highest development in the university. The basic function of a university should be to develop the ability to think: training to do should, at the university level, be added to education, never, as it so often is, substituted for it.

Culture is the organized behaviour of a society. It is dynamic, not static, and is continually being reshaped. Humanism is the accumulated thinking of mankind while science is the classification of observed facts— the advancement of knowledge. Science is, therefore, part of both culture and humanism and the humanist must be able to interpret science for the benefit of society.

It is society—not science—which creates technology—the technology on which our modern North American culture is based. The humanist must be able to control the advances in technology and must be able to distinguish between science and technology. The technician is trained: the humanist is educated. Sometimes one individual may be both, but not very often. There is no long-term shortage of trained individuals; there is a critical shortage of educated ones.

Our system of schools, colleges, and universities is tending more and more towards standardization. In a material culture, dominated by advertising and catchwords, the pressure towards standardization is enormous. The most important task we have before us is to resist this tendency, to appreciate the biological fact that no two individuals have equal potentialities, that only some, and a relatively small proportion at that, can be educated as constructive thinkers. It is equally important to realize that the potentialities for this are not defined by social position or wealth, but are widely distributed. They are so scarce that no effort must be spared, either by the family or by the society, in developing these potentialities to the full. This biological heritage of mankind is a debt from the past which must be paid in the future.

Any plans for the redemption of this debt must take into account the biological factors in the human organism; these factors cannot be ignored in any properly planned society. The society must remember:

First, that man's internal environment is under only partial mental control. Many of his actions are the result of instinct or the secretions of his endocrine system rather than the result of reason. The ability of the mind to control these reactions varies from individual to individual; although their control may be helped by the medical sciences, it probably can never be complete in any one individual. The law of the society must always bear this in mind.

Second, that although potentialities can be developed and most individuals trained in some art or craft, only a very few can be really educated to be constructive thinkers. Owing to the highly mixed genetic constitution of mankind, these potential thinkers may be found in any stratum of the society. The society, therefore, must seek them out and provide them with the opportunities to develop their particular talents.

Third, that still far too little attention is being paid to the human biological sciences, those which concern man as an organism. An infinite amount of effort has been expended in improving his material environment, much has been done in improving and maintaining his physical condition, but very little has been done in attempting to understand the factors influencing man's mentality or the effect of his endocrine system on his internal environment or the nature of his instincts. Very little more has been done in investigating methods whereby both education and training can be applied with the best prospects of developing an individual's potential talents. It is self-evident that no two individuals are biologically equal; it is equally evident that a society in which all are legally equal requires the fullest use of the abilities of all its members. Such a society can evolve and create its own distinctive culture only when it understands and can influence the biology of all within it. This is true of all societies everywhere. It is particularly important, however, to the Canadian society with its almost unlimited possibilities for the future.

A. S. P. WOODHOUSE

YOU WILL PERHAPS remember Carlyle's remark when told that the lady from Boston had decided to accept the universe: "Gad," said Carlyle, "she'd better!"

This is a sound principle, I think, for the humanist when confronted by the findings of natural science; and it would be idle, as well as impertinent, for me to question the description of man just given by my colleague—so far as it goes. Only, I would remark that, like other descriptions, it is conditioned by the describer's point of view and the order of facts to which he directs attention. The natural scientist must reduce the human problem to corporeal terms on his way to an acceptable solution. The humanist distrusts reductionism in every form and insists on giving to all the developed phenomena their full weight, whether he can explain their origin or not. Biology presents us with an account of the animal, man; and this, like any other set of established facts, we ignore at our peril. It gives us the physical bases of man's achievements and warns us against certain elementary mistakes about human nature. It is not, by itself, capable of setting man's goal or of telling him how to reach it. Biology informs us that man's power to think and to invent depends on a development of the frontal lobes which distinguishes him from all other animals, and this discovery is assuredly a triumph of the aforesaid frontal lobes. Or (to speak a little cheerfully) who but the biologist found the gene in genius?—another triumph! But do we, in fact, understand, appreciate, evaluate the works of genius—of Homer or Shakespeare, Michelangelo or Mozart, Plato or Kant, Newton or Darwin or Einstein—the better for this knowledge? The answer is that we do not. Yet the perpetuation of culture depends in part on the ability to understand, appreciate, evaluate the work of genius.

No doubt a high degree of social organization can be achieved without this ability to understand. I suppose the statement that man "is still not really a social animal" means, among other things, that the human individual is not so completely submerged in the life of his community as are those efficient totalitarians, the ant and the bee, and that indeed he resists submergence. But is the implication that he should be submerged and (if he develops satisfactorily) he will be? We must not confuse his resistance with those anti-social attitudes and actions which spring from what Huxley called the impulse of unlimited self-assertion inherited from our ancestors of the jungle, which (he added) was the truth embodied in the doctrine of original sin. Certainly we must let the ape and tiger in us die, but this is something very different from the suppression or submergence of individuality. Society and the individual (whatever tensions arise between them) are interdependent entities; and your true proponent of individuality is not the bank robber or some other anti-social being, but (shall we say?) John Stuart Mill writing the essay on Liberty.

Aristotle called man "a political animal," and observed that if human beings came together for the sake of mere life, they remained together for the sake of the good life. In the hard logic of facts, life is prior to the good life; hence the fundamental importance of biology and of those social sciences most closely related to it. But if we expect from them an adequate description of the good life, we are likely to be either disappointed or deluded.

Natural science tells us (if experience has not got there first) that man is subject to the needs and drives of the animal kingdom, and, like its other denizens, is under what Bentham called those "two sovereign masters, pain and pleasure." But to complete the picture one must consider the restraints which man has put upon these natural drives, and the refinement of them which he has conceived and sometimes attained; and one must remember that for civilized man the pleasurable and the painful are far from being the simple categories which the supreme reductionist Bentham assumed. Nor, despite its immense importance in his evolution, must we confine our attention to man's ability, under the goad of necessity, to think and to invent. We must add that later acquisition, his capacity for disinterested enquiry, and for the peculiar form of disinterested invention which we call art; and to intellect in its various activities we must add the whole realm of feeling, of sensibility, which may become in its turn an object of thought and is certainly susceptible of cultivation.

If what we are talking about is civilization, the conservation and progression of man's cultural achievement, or what Aristotle called the good life, I think we must add to the principles and programme of the scientist a definition supplied by the humanist. "Civilization," said Matthew Arnold, "is the humanization of man in society. Man is civilized when the whole body of society comes to live with a life worthy to be called human and corresponding to man's true aspirations and powers."

Every word of this definition is packed with meaning. The word "humanization" recognizes in man (however it came there) a specifically human and rational self which can be fostered and fortified, as well as an animal and instinctive self which must be lived with and brought under wise control. In its context, the phrase "in society" recognizes that man is indeed a social being, not simply because for him society is the necessary condition for mere life (as of course it is) but because it is also the indispensable condition of the good life, of the humanization of man. The concern for "the whole body of society," and (under a proper understanding of the term) for equality, is no sentimental equalitarianism. The humanist knows that men are born neither free nor equal. All he demands is that individuals should never be regarded merely as means to ends, but that all should enjoy equal opportunity for pursuit of the good life, which (given this opportunity) can be achieved under very different conditions and in a variety of forms. He is far too sensible of the variety in human nature to expect uniformity, and he is the resolute opponent of levelling down, that is, of reductionism transferred from theory to practice. Nor is there anything vague about the phrase "man's true aspirations and powers." It connotes all that we mean by morality, positive and dynamic, including our duty to society, all that we mean by knowledge and the exercise of intellect, and by the response to beauty, and (though this may startle us) it includes good manners and social grace. To achieve these things in any significant degree is to progress towards "a life worthy to be called human," or, in other words, towards the good life.

For the race it has been, and presumably always will be, a long and arduous adventure, with many an obstacle, internal and external, many a hazard, and at times serious set-backs. A recurrent problem is posed by the numbers demanding their share in a better life, as they conceive it —often crudely enough. When the barbarian hordes overran the Roman empire it took a thousand years to recover the lost ground, but there is reassurance in the fact that the new culture, as it gradually emerged, contained elements that the old had never known. For the last hundred and fifty years, the problem of numbers in its modern form has been growing ever more acute, so that Dean Inge was moved to remark that while ancient civilizations fell before barbarians from without, we breed our own. Ironical as it sounds, there is some advantage here, if our educational system can keep pace with the demand made upon it and our educational philosophy can escape the dual dangers of inflexibility and debasement. But now the problem is complicated by a rapidly changing world scene, with the peoples of two more continents asserting their claims, and a power of the first magnitude acting as a focus for much that is opposed to our most cherished values.

What the outcome will be, who can say? Technology offers no solution. It is a good servant—that is all. Dr. Cameron was concerned to disengage science from its presumptuous offspring; and this is a step towards clarification. But does science by itself hold the ultimate solution? One may be permitted to doubt it.

The solution, whatever it be, must be commensurate with the problem; and it must be sought in the whole nature of man and in the light of his total history—not in what Arnold called the "power of intellect and knowledge" alone, much less in that power when limited to the field and method of natural science. So much, indeed, science itself might teach us, and (I believe) does; for surely a first principle of science must be to take into account all the relevant data, and a second, to apply to the data, whatever they may be, a method competent to deal with them. Not less than the scientist, the humanist is committed to the primacy of knowledge and of informed, clear and consistent thought. But he recognizes that civilization depends on much besides: depends on the cultivation of the whole range of human sensibility—of imaginative sympathy, of admiration, of response to the good and the beautiful—and on the cultivation of those principles of human dignity and moral obligation without which civilized society is impossible. And he insists that on these subjects also informed, clear and consistent thinking must be brought to bear.

Thus he believes that education is the principal means of preserving and spreading, if not of advancing, civilization. Properly understood, education is uncompromisingly intellectual in method, and stands at the farthest possible remove from mere indoctrination and from mere adjustment to the *mores* of the herd. For it consists in the critical assimilation of knowledge, that is, of the facts of human existence—from the biological and economic facts on which man's life depends, to the moral, political and aesthetic, which pertain to the good life; but really to assimilate what pertains to the good life is to add to knowledge and intellectual training some extension or refinement of understanding, sympathy, and sensibility.

Advances in civilization, as Dr. Cameron has emphasized, must always be the work of the specially gifted few; and the degree of their dependence on education will vary greatly in different areas and different individuals. But of their indirect dependence on the prevailing state and temper of civilization in their day we may be certain, and this in turn depends directly on education. Such education should in its character reflect, and in its operation fortify, two of the distinguishing marks of a civilized society: permanence and progress. It should enable us to take secure possession of whatever is best in our inheritance from the past and at the same time to welcome and incorporate whatever advances our own age can effect. Thus the inheritance is handed on unimpaired and (in the degree possible) augmented.

If one is asked (as I have been by our chairman) to suggest the bearing of all this on Canada today, one must insist that societies do not differ in the principles of their well-being, but only in their circumstances and in the particular problems to which these circumstances give rise. In two respects Canada's circumstances are fortunate, though each gives rise to its own problems. She has great natural resources and a great industrial potential; in other words, she has the means of mere life, which is a necessary condition for the good life, but from which the good life does not of itself follow. And she is heir to two of the world's greatest cultural achievements and to the languages in which their deepest insights and highest aspirations have been couched. On these her hold has been tenacious indeed, but at best partial. She is not in any real sense a bi-lingual and bi-cultural nation; she is merely a nation with two languages and two separate cultures. As to her augmenting of her cultural inheritance, it is no doubt natural that this should have occurred (so far as it has occurred at all) mainly at the level nearest to the material and the practical, namely, in the field of politics broadly conceived. In art and letters (I speak particularly of my own racial group) her performance so far has been, with a few honourable exceptions, meagre and disappointing. But here the power, and hence the responsibility, of the community is limited. It cannot produce genius at will. It can only, by means of education, foster in itself an appreciation of true genius, past, present or yet to come.

ROBERT ELIE

FAUT-IL S'INTERROGER sur l'homme chaque fois que l'on en vient à parler de civilisation, d'humanisme, de tradition ? Oh ! l'encombrant animal qui déborde toujours la définition que le savant a si patiemment élaborée. Que n'est-il prévisible comme la méduse ! Ce solitaire qui feint de rechercher son semblable est la plus féroce bête de la jungle. Son sourire, ses caresses, ses discours ne sont que pièges. J'en appelle à l'histoire et je l'accuse. Mais qui accuse l'homme ? N'est-ce pas encore l'homme ? Vous le voyez : il bouge, il esquisse un geste; il a vraiment l'air d'un animal. Bien plus, il parle, il dit qu'il fait beau, et c'est vrai qu'il fait beau; il serait donc doué de ce qu'on est convenu d'appeler intelligence.

L'homme est un animal intelligent. Ce ne sont encore que des mots et l'homme dit à l'homme : tu es le plus vulnérable des animaux et tes petites idées s'évanouissent comme bulles de savon dans les espaces infinis. Et le procès reprend de plus belle, mais procès inutile puisqu'il n'y a pas de juge. Je veux bien que l'homme s'interroge, c'est même par là qu'il est grand, mais je me méfie dès qu'il se met à s'accuser.

Je ne prévoyais pas que le débat s'élèverait à d'aussi vertigineuses hauteurs. L'évolution est une chose toute neuve pour moi, et je crois y trouver des merveilles. C'est l'espérance rajeunie et un tel regain de confiance que l'homme pourrait enfin abandonner à plus grand que lui un procès qu'il s'entête à soutenir. Si un passé de millions d'années qui se dégage enfin du brouillard, ne saurait encore garantir l'avenir, du moins je croyais qu'un Teilhard de Chardin avait précisé davantage, grâce à l'évolution, le sens de l'aventure humaine. N'est-il pas exaltant le destin de cet animal surgi de la matière qui, un jour, franchit le seuil de la réflexion pour reconnaître le monde, dire le réel, s'étonner de ses premières pensées, fraîches et transparentes comme la lumière du matin ? « L'homme sait qu'il sait », et personne d'autre sur cette planète ne possède ce prodigieux secret. Laissez-moi croire que la naissance de la conscience, avec ses inquiétants mystères, fut une étape décisive de la vie, et non pas un accident qui aurait fait de l'homme un animal infirme.

Mais ce serait trop de candeur que de vouloir suivre mes savants collègues sur des voies dont je ne fais que soupçonner l'existence et d'invoquer le témoignage d'un maître dont le langage me dépasse. Mais je ne puis m'empêcher de protester contre l'opposition que l'on s'empresse d'établir entre conscience et vie, entre chair et esprit. Si je ne puis me réclamer des savants et des philosophes, du moins de très chers poètes, de fraternels peintres me disent que l'accord est possible. Et je me tournerai avec autant d'assurance, au risque de paraître peu évolué, vers les plus humbles des hommes et des femmes qui ont traversé ma vie pour affirmer qu'il survient des moments d'exaltation de tout l'être qui me font rejeter cette image d'un homme qui ne serait qu'un animal de proie un peu plus sournois que les autres.

La conscience engendre l'amour. L'homme seul sait qu'il sait, et seul, il aime aimer. Cet animal n'a vraiment pas fini de nous étonner et je me refuse à croire que ce génie de l'amour soit le privilège de quelques rares esprits, exceptions qui confirmeraient la règle de la plus abjecte bestialité. Un tel aristocratisme me paraît aveugle; l'histoire ne vient-elle pas de nous rappeler quelles jolies bêtes font ceux qui se croient supérieurs ?

Il y a la précieuse communication du sage, mais nous pouvons encore interroger ces humbles regards qui ont la profondeur du bonheur de vivre. D'une œuvre d'une grande érudition, ce « traité de métaphysique » où Jean Wahl cherche à son tour à préciser le sens de l'aventure humaine, un critique, Paul Ricœur, détachait récemment ces lignes admirables : « Pour certains d'entre nous, c'est de quelques événements de 1939-40 qu'il faut partir pour construire, aujourd'hui, en nous l'idée de l'homme, en pensant à ce que l'homme a souffert et à la conscience qu'il a prise, alors, qu'il y a des choses pour lesquelles il fallait souffrir. Nous avons vu que l'homme peut supporter beaucoup plus qu'il ne paraît : que l'homme, tout en disant : plutôt la mort que..., et ayant toujours sa propre mort comme une arme dans la main, ne médite pas sur la mort, que l'homme est une créature d'espoir. Et par là même une créature de courage ».

Les voies de la conscience sont innombrables, et ses manifestations, diverses. Déjà, les dessins de notre ancêtre des cavernes sont œuvre d'une pénétrante intuition. Ce trait qui définit une forme avec tant d'assurance est signe de la domination de l'homme sur l'univers, mais l'œuvre ne vaut pas uniquement comme reproduction des choses : il a sa valeur propre que l'on commence seulement à comprendre. Bien au-delà de l'image, au-delà des apparences de l'œuvre d'art, éclate une lumière qui vient de l'intelligence, une pensée se manifeste. Si l'œuvre, en tant que représentation du monde extérieur, indique que l'homme sait; cette même œuvre, en tant qu'objet de pensée, indique que l'homme sait qu'il sait.

Cette voie de la poésie, qui conduit aussi à la sagesse, est ouverte à tous, et c'est pour en avoir méconnu la valeur et pour n'avoir vu dans l'art que le reflet de la nature que l'on a vu apparaître ce type d'intellectuel désincarné et myope qui refuse l'accès à la conscience à la foule qui ne peut suivre la voie de l'érudition.

J'en conclus que l'humanisme qu'il nous faut inventer doit tenir compte de l'enseignement de la vie et de ces créations spontanées qui en sont la manifestation.

Auprès du témoignage de l'art, des paroles lumineuses de la poésie, il y a ces voies de la recherche scientifique, dont d'autres pourraient parler mieux que moi, cette longue et patiente quête de vérité qui a permis à l'homme d'échapper aux sortilèges de la magie, de mieux dominer le monde, de se libérer de tâches avilissantes afin de poursuivre avec plus de vigueur son projet de conscience.

Je ne saurais dire si le pouvoir de réflexion s'accroît dans l'homme, mais il me paraît évident que son exercice n'est plus réservé à de rares individus, ce qui, à des époques moins démocratiques que la nôtre, a pu conduire des penseurs d'ordinaire médiocres à croire qu'ils étaient des créatures exceptionnelles. Diafoirus n'a jamais imaginé que le sculpteur anonyme d'une invisible statue de la cathédrale de Chartres le devançait sur les voies de la conscience.

Si toujours plus d'hommes accèdent à la conscience, n'est-ce pas une preuve qu'il y a communication entre eux, que les créateurs réussissent à transmettre aux autres leurs découvertes, que tradition n'est pas un vain mot, bref que l'homme est un animal social. Non seulement les découvertes se communiquent — même le sourire de la mère est pour l'enfant comme une profonde parole — mais je suis convaincu que le besoin de communiquer avec l'autre est la cause des découvertes. Quel savant ne s'empresse de communiquer les résultats de ses travaux, quel poète hésiterait à publier cette parole de vie qui lui paraît éclairer l'étonnant destin d'une créature d'espoir et de courage ? Et je crois qu'il y a là plus d'amour que de vanité.

Que la joie et la douleur soient les mobiles de nos actes, est-ce que cela prouve que l'homme n'est qu'une bête de proie ? Il y a la joie de donner et la douleur de ne pouvoir tout donner, la joie de vivre et la douleur de ne pouvoir atteindre à l'absolu que l'on désire : ce sont là des sentiments qu'aucune bête ne connaît et qui témoignent de la grandeur du projet d'homme.

L'aventure de la conscience est bien l'affaire de tous. Tout esprit créateur finit par rejoindre le plus humble de ses frères, et le plus humble arrive à comprendre quelque chose de l'œuvre du créateur; comprendre, c'est-à-dire prendre avec soi, et se servir d'une chose, c'est commencer à la comprendre. Ce tableau du Titien, aussi vivant qu'au premier jour, cette fugue de Bach, cette tragédie de Shakespeare attirent encore dans leurs profondeurs des hommes qui en communiqueront la lumière à d'autres.

La communication est encore plus généreuse, le réseau des influences plus étendu. Teilhard de Chardin nous demande « d'observer un peu mieux autour de nous : ce déluge soudain de cérébralité; cette invasion biologique d'un type animal nouveau qui élimine ou asservit graduellement toute forme de vie qui n'est pas humaine; cette marée irrésistible de champs et d'usines; cet immense édifice grandissant de matière et d'idées... Tous ces signes, que nous regardons, à longueur de journées, sans essayer de comprendre, ne nous crient-ils pas que sur Terre quelque chose a « planétairement » changé ? »

Les esprits chagrins se plaisent à souligner les plaies du paysage urbain et son désordre. Depuis dix ans, nous voyons grandir nos villes et il est vrai qu'elles ne sont pas belles dans leur âge ingrat, mais qui resterait insensible aux forces prodigieuses qui animent ces corps encore informes, poursuivant leur œuvre sans se préoccuper des doctrinaires sentimentaux. Ces forces, moins aveugles qu'on ne les croit, créent leur ordre. Déjà, des perspectives s'ouvrent dans nos paysages urbains, une architecture nouvelle donne forme à ces forces que le savant a libérées, et elle les rend intelligibles. Du sommet de l'un des grands buildings de New-York, ne voyons-nous pas se parfaire, d'année en année, une immense pyramide qui aura sur les monuments anciens l'immense avantage d'être un lieu de travail pour les vivants, et non pas un somptueux tombeau.

Ce qui se passe dans le silence du laboratoire, dans l'atelier du peintre ou le cabinet de l'écrivain intéresse tous les hommes. Cette poussée de conscience est d'ailleurs irrésistible, même si elle rencontre bien des obstacles, et la vie se charge de réfuter les prophètes de malheur qui refusent de regarder en avant. Ils nous ont dit hier, et peut-être l'ont-ils répété ce matin, que le texte doit céder la place à l'image, et pourtant qui aurait imaginé il y a seulement vingt ans que l'on publierait les œuvres de Platon et de Keats, de Pascal et de Flaubert dans des collections populaires qui paraissaient réservées aux auteurs de romans policiers ?

Je souhaite que le nouvel humanisme craigne moins que celui que l'on nous a proposé de suivre la science dans ses découvertes, car elle libère des forces créatrices de vie.

L'ampleur du projet effraye encore. Les sociétés hésitent devant leur avenir comme si elles craignaient de perdre les précieuses acquisitions du passé. Mais rien ne me paraît plus malsain que de rêver à un nouveau

Moyen-âge ou à un retour à quelque forme ancienne de civilisation. Je n'ai aucun goût pour les restaurations qui momifient le passé, et je crois que les créateurs seuls sont respectueux de la tradition.

Toute révolution dans l'histoire et dans la vie apparaît au départ comme réaction. Si le mot révolution semble excessif, disons toute invention, tout renouvellement. Il y a d'abord rupture avec de mauvaises habitudes qui engendrent la répétition. Les traditions ne sont plus alors que conventions, et si tyranniques que la moindre tentative de renouvellement ou d'expression personnelle se trouve condamnée. Et pourtant, le présent, qui ne propose jamais le même équilibre de forces qu'hier, exige pour être vécu et dépassé, pour que demain devienne présent et que l'homme ne s'enlise pas dans un passé qui n'est qu'un souvenir nécessairement infidèle, ce présent exige une réponse nouvelle, assez compréhensive pour qu'elle nous permette d'aller au-delà et d'épouser ainsi le mouvement de la vie, exactement comme l'ont fait nos grands aînés qui ont eu la générosité d'accueillir le présent.

Et voilà le point de rencontre, de ressourcement, de renouement avec la tradition. Le créateur se laisse bientôt prendre au jeu de la vie; il ne s'agit plus de réaction, mais d'action. Seuls les esprits médiocres continueront à protester quand d'autres auront rejoint la vie, et ils seront seuls à refuser le proche avenir.

Le créateur s'étonne que la réponse qu'il offre à un présent nouveau soit une affirmation de l'homme qui en rejoint d'autres qui furent proposées dans le passé. Mais il ne s'agit plus du passé, car tout redevient présent, les générations renouent entre elles, la tradition est enfin continuée sans trahison. Proust retrouve en Montaigne et Saint-Simon des frères à peine plus âgés, Renoir interroge les grands Vénitiens et la Bible paraît familière à Claudel. Ainsi, toutes les réponses que les hommes ont offertes à l'interrogation du présent forment, dans leur diversité, une seule affirmation de l'homme devant l'avenir qui lui propose l'infini et l'absolu.

«Ce n'était donc pas assez de dire, comme nous l'avons fait, écrit Teilhard, qu'en devenant consciente d'elle-même au fond de nous-mêmes, l'Evolution n'a qu'à se regarder au miroir pour s'apercevoir jusque dans ses profondeurs, et pour se déchiffrer Elle devient par surcroît libre de disposer d'elle-même, — de se donner ou de se refuser. Non seulement nous lisons dans nos moindres actes le secret de ses démarches. Mais, pour une part élémentaire, *nous la tenons dans nos mains* : responsables de son passé devant son avenir. »

« C'est tout le problème de l'Action », ajoute Teilhard. C'est le problème de la fidélité au présent, où se rencontrent le passé et l'avenir.

C'était là mon sujet, mais j'ai suivi mes collègues dans des voies plus ambitieuses. C'est ce présent qu'il me semble urgent d'interroger, car c'est par une fidélité de tous les instants que l'on participe à cette aventure de la conscience qui est l'affaire de tous, propos de connaissance et d'amour qui est l'essentiel de l'humanisme.

J'aurais mieux fait d'explorer le petit espace que je puis prétendre connaître, mais peut-être n'est-il pas inutile de répéter que l'homme a un avenir qu'appelle tout un riche passé, c'est-à-dire un présent où rien n'est perdu de la patiente recherche des générations, où rien ne permet de croire que l'aventure humaine s'achève.

ROY DANIELLS

   AFTER THREE SUCH PAPERS AS THESE, which have been so cogent and yet so suggestive, what remains to be said? They have set forth the claims of body, mind and spirit—if that ancient, improbable but convenient tripartite concept of man can be used in such a company as this. And the claims of body, mind and spirit have been used as starting posts and points of control for the three arguments. What, then, can possibly remain to be said? I can think only of the feeding of the five thousand, as recorded in the Gospel; after which miraculous satisfaction of those who, out of a desire for truth, had ventured far from home, there seemed nothing left to do. And yet, there was something, a small task. "Gather up the fragments that remain that nothing be lost." I can see one of the disciples, Doubting Thomas or James the Less perhaps, fulfilling that small task.

One such fragment is the observation that certain patterns of opposed or complementary elements have presented themselves. Taking these pretty much as they occurred, we find: instinct versus reason; internal environment as against external environment; psychology and physiology; science and humanism; education and training; society and the individual; biological drives and humane restraints and refinements; material invention and disinterested enquiry; pragmatic rationalism and sensibility to the arts; man as a beast of prey and man as one who loves and voluntarily suffers; the way of poetry and the way of erudition.

These pairings present their own problem. Do they exist thus in the nature of things, or do our Western minds, provided by their education with opposing categories, demand a chess-board with black and white pieces, of fixed powers? As time goes on and modes of thought change, shall we not give more consideration to non-Aristotelian logic, to degrees of probability, to "possibility" as a category of equal value with "necessity"? Our mechanisms of thought must be prepared to deal with opposites; with complementary pairs; with paradoxical pairs; and with related pairs which, like Conservatism and Social Credit, will not admit relationship.

While the arguments put forth this afternoon are clear to this audience, with its special training, they may not be clear in the near future and to a more heterogeneous group of hearers. There is need for definitions and a scrutiny of meanings. I look casually into the sort of paperback that young students are reading and find statements such as this one on the subject of Time: "It is a figment of my thinking. That as such it might some day put an end to my thinking, as some believe, is beyond my comprehension. Even the old myth makes Kronos devour only his own children, not his begetter." Such a passage passes from scientific, to humanistic, to poetic modes and invites attention from someone concerned with semantics. The great debate, to which the cogent addresses we have heard are contributions, must go on. The subject of this symposium is our debt to the future. One of our obligations is to see that the debate goes on into the future in the best possible and (for the future) the most comprehensible style.

But what really is the debate about, this great debate which is the only real substance of education? I venture to guess that it is teleological and concerns our ends. The three papers we have had the privilege of hearing have given us a sense of sweeping through history, of continuity and causal relationships. They have given us a sense of ends, whether these are instinctive or rational, implicit at the beginning or only to be achieved by trial and error. This is not inconsistent with the oppositions and pairs of contraries mentioned. Bifurcation, as Hegel implied and as any infant discovers when he gets up on his two feet, permits progress.

The debate which the future deserves must go on, but with more attention to articulating a mechanism of thought which is neither too loose nor too cumbersome; to finding a set of concepts, or images (as the poet might prefer it) which the common man like myself may use. Let us remind ourselves how Matthew Arnold laboured to produce simple formulas of debate which became useful and of widespread value. In this debate about purposes and ends, therefore, we need humility as a path to simplicity, the simplicity of a Newton, an Einstein, a Wordsworth, and of some of our contemporaries here in Canada, whom we could name if it were not thought to be invidious.

The debate must continue, with its semantic problems recognized, with the knowledge that some of its cruxes may be verbal, that the human mind is the mechanism by which scientist, humanist, and artist alike work, and is capable of its own reconciliations.

The debate must go on with the object of clarifying ends, for nothing cuts more quickly through the felted texture of Canadian pragmatism than such a debate. There are, indeed, no lack of proposals as to the great ends we are alleged to be pursuing. Reason, common sense, tradition, authority, intuition, experiment, revelation: each supplies or purports to supply an answer. And no one answer (if I may at some point indicate a personal impression) no one answer is enough. The debate must go on, for if we stand at the junction of immense vistas, it is with the knowledge that we cannot stand in perpetual wonder, *we* must go on.

The debate will proceed into the future and we have an obligation to the future to conduct it with the utmost scrupulousness and with disinterestedness and earnestness—to use two Victorian words of known connotation. With a humble sense of our limitations we must pursue an ideal simple enough to be generally comprehended, an ideal which maintains the sort of connections you have just seen constructed between biology which is basic, the values of humanism which are indispensable, and the insights of the poet without which "the people perish."

One simple and effective method used in the past to clarify such concepts is the projection of an ideal man: the saint, the martyr, the prince, the courtier, the gentleman, the wayfaring Christian, the man of sensibility, the utilitarian—as the needs of each oncoming age seemed to demand. Each of these ideals was powerful in its own right for a longer or shorter duration. But since Arnold's man of culture, where have we had such another embodiment?

I would not venture to define or project an image of the ideal young man in Canada, whom we are now educating and by whom our debt to the future will ultimately be paid. But his outlines take shape dimly in the mind. Friends pass in memory and each contributes a lineament, a moral or intellectual trait. He is not a chimaera. We have seen him. I think that for a year I lived with him in the person of a young mathematician born and educated not many miles from here. But his appearances in our society are fitful. Our duty to the future is by the great debate to evoke him, make him visible, identify him, give him form and substance in the popular mind. It is not impossible. These things have been done. He is not an Isaiah, a Socrates, a Boethius, an Augustine, a Pascal or a Wordsworth. But something other, more modest, for the future age more viable, and inevitably more our own.

LET US LOOK TO OUR HUMAN RESOURCES

F. H. UNDERHILL, F.R.S.C., *Chairman*
J. K. W. FERGUSON, F.R.S.C.—L.-P. DUGAL, F.R.S.C.
W. B. LEWIS, F.R.S.C.

F. H. UNDERHILL

WE HAVE NOW COME to the last of this symposium. The final session should be a kind of summing up, and the committee has suggested, as topics for discussion, "the quality of our national life; the efficient use of manpower; the contribution of education in the schools, in the universities, and at the adult level; the influence of mass media; our philosophy of life." Dr. J. K. W. Ferguson, who is director of the Connaught Medical Research Laboratories, University of Toronto, is going to talk on health and our human resources; Professor L.-P. Dugal, chairman of the Department of Biology, University of Ottawa, will discuss our human resources and the aims and purposes of education; and finally Dr. Lewis, Vice-President, Research and Development, Atomic Energy of Canada Ltd., of Chalk River, Ont., will speak on the efficient use of manpower.

J. K. W. FERGUSON

EARLIER IN THIS SYMPOSIUM the dismal problems of overpopulation were thoroughly discussed. Barring an atomic catastrophe, it seems unlikely that there will be any scarcity of human resources in times to come. It is not, however, my intention to discuss the numerical aspect of human resources. Instead, I shall attempt to assess some factors affecting the quality of human life. One of these is health.

The physical health of a nation is indicated to some extent by the life-expectancy of the population. The life-expectancy of individuals in the Western world has been increasing steadily for at least a hundred years. The various medical arts and sciences can, I think, accept most of the credit for the increases in longevity of the last hundred years. Longevity is not, however, the same thing as, or necessarily a cause of, overpopulation. Medical science may take the credit for one without assuming the blame for the other. It almost certainly had little to do with the increase in population in Europe during the eighteenth and early nineteenth centuries, or perhaps even with the population increases in Asia and Africa during the present century. In the latter situations, I suspect that improvements in the production and distribution of supplies of food have been much more important.

If it is conceded that the increased life-expectancy, as distinct from increased population, in the Western countries has been due largely to the application of medical knowledge, we must conclude that women have benefited much more than have men. The life-expectancy of both sexes has increased, but that of women much more than that of men. It is hard to see why this should be so. The control of infectious diseases by sanitation, immunization, and more recently by chemotherapy and by antibiotics has probably been responsible in large measure for the great reduction in mortality of infants and children of both sexes and to a lesser extent in the less dramatic reduction in mortality among adults of both sexes.

The discrepancy in life-expectancy of men and women, in Canada, is about four years, and if we accept published figures at face value, it seems to be a little higher in Canada than elsewhere. But the important point is that this discrepancy in the life-expectancy of men and women is increasing. A number of questions present themselves immediately. Why are women now living longer than men? It was not always so. Is the difference in life-expectancy going to increase further? Are women now having too easy a life? Are they carrying far too little of the load of modern life? Is it time to come to the rescue of the overburdened male? These are intriguing questions, but ones which I feel no great obligation to answer just now. So let us consider another feature of the present which has implications for the future.

Medical progress has had profound effects on certain attitudes. Faith in medical experts has increased enormously. Demands for medical services have increased accordingly. Such demands are both gratifying and embarrassing for it is hard to see how maximum service can be allocated freely and impartially, and yet be paid for. Our equalitarian spirit rebels at the thought that some people may get better attention than others. Various means will doubtless be found which will at least provide more uniform medical service for the bulk of our people. Will it be better service on the whole? It can be if the demands of all concerned are reasonable.

In our time we have seen a remarkable approach to socialistic uniformity of income in our population. With political power widely dispersed, as it is in our democracy, there is every reason to believe that the trend to uniformity of income will increase, and that the prosperity which is forecast for the next few decades will be distributed with even greater uniformity throughout the population.

Fear has been expressed that the uniformity will be so great that it will deserve the epithet of dull conformity. As far as I can see, the threat is not very serious. For better or for worse, it seems likely that non-conformity and unpredictability will continue to be prominent features of human life. To put it pessimistically, there is every reason to believe that many people will continue to be as crazy as ever for a long time to come. Madness in a literal sense continues to be very important as a source of unpredictability in modern life and so does crime. The thesis very popular until recently among social scientists that poverty is the major cause of crime seems hard to support in these days of prosperity. The hope that delinquency will decrease automatically as a result of greater prosperity seems to be over-optimistic. Neither mental illness nor moral delinquency has yielded appreciably to the attacks of science or the benefits of prosperity. The modern tendency to classify certain moral and emotional aberrations as diseases has accomplished little as yet in the way of prevention or control of such aberrations.

Let me say quickly that I am not seriously opposed to labelling as diseases certain kinds of behaviour which can be regarded either as disease or as delinquency or as a mixture of the two. Calling such conditions diseases sheds no light on their nature but does serve a useful purpose. It makes them acceptable as subjects of polite conversation and enlists support for objective investigation. Mental and moral aberrations constitute the greatest challenge facing the medical and social sciences. It is worthy of note that the act establishing the Canada Council stated as its purpose the encouragement of the "arts, humanities and social sciences." One hardly knows whether to be pleased at this recognition of the importance of the social sciences or concerned that they have been so treated, almost as an afterthought.

Another attitude which has changed drastically during the last fifty or hundred years, and which may be related to increased life-expectancy, has been the public attitude towards education. By education I do not mean something very sublime or rare, but merely education in the colloquial sense, which most of us understand as organized instruction.

This kind of education is something very much in demand by the public. Nearly everybody is in favour of it and wants more of it, if not for himself, then for his children. Only a few recalcitrant religious groups and some reluctant children retain a conviction that the whole thing can be overdone. The power of the popular urge for education was well exemplified by the serious effort which was made in the United States shortly after World War II to provide a college education for everyone. This widespread aspiration should not, I think, be dismissed with the cynical comment that a B.A. diploma with every birth certificate would be much simpler. A better answer, I think, is the question: "Who really wants a B.A. diploma?" Perhaps the reluctant children have some rights.

For many years a large proportion of our secondary school pupils have left school before they graduate. A figure of 40 per cent is sometimes quoted. This rate of desertion or depletion is regarded as deplorable. On the other hand, we are hearing more now about the evil of allowing pupils to remain in school if they will not work. Such pupils debase the quality of education which can be given to those who are eager to learn. It seems that we must soon make up our minds as to which evil is more deplorable. The inescapable fact of a shortage of teachers for secondary schools and colleges which cannot be overcome in the next ten years or longer may settle the question for us. It will probably be necessary to select more carefully those who may receive the privilege of secondary education and of university education. Must we accept this as a regrettable expedient, or can we adopt it as a desirable principle? If the latter, do we thereby condone discrimination and privilege, or even establish a form of undesirable discrimination as an ideal? I think not, provided that those who are denied schooling can be given education suitable to their needs by alternative means.

Perhaps we should recognize more clearly that some kinds of education require schools and schooling. Other kinds do not. Many years ago, it was customary to classify subjects of education as "cultural" or "disciplinary." No doubt these terms fell out of fashion partly as a result of the different meanings which these words carry for different people. Perhaps they could be revived with profit and given acceptable meanings.

Suppose that we define a disciplinary subject as any science, art or skill which requires intensive personal instruction and criticism on the part of the teacher and critical performance and effort on the part of the student. Suppose that we define cultural subjects as ones capable of transmission to large groups by attractive presentation, and requiring little performance on the part of the student beyond a certain degree of interest and attention.

Someone may (and indeed should) object immediately that these classifications are quantitative, not qualitative. They do not distinguish music from arithmetic. They merely distinguish listening to music from learning to play music, or to write music. Quite so! It is not the name of a subject which determines whether it is cultural or disciplinary, but what one learns to do with it.

At this point someone else may object that such a definition associates cultural with elementary, shallow and ineffectual. My retort is that the association can also be with more respectable words like broad, comprehensive, sympathetic and perceptive, all good things to be!

Having made the attempt, I admit failure. Disciplinary stands. Everyone knows what that means. But cultural—no—it is too late to make an honest—word—out of that one. It is too highly charged with strong emotion and vague meanings. Let us try another word to take its place so that we can get on with the discussion. Let us use "evocative" —the calling forth of wider interests and desirable attitudes.

If you will grant me the terms disciplinary and evocative as distinguishing, for some practical purposes, two kinds of education, my propositions are simple. (1) Disciplinary education needs schools and schooling; evocative education does not. (2) All disciplinary education need not be acquired in conventional schools of secondary and higher level.

At this point, an interesting question obtrudes itself and threatens to cause a digression. Which is more important, evocative or disciplinary education? Let us deal with this one quickly. Of all the education and knowledge which even learned people use all day and every day, what proportion is broad and casual, not to say shallow, and what proportion is precise and highly disciplined? Surely no answer is required! We may now return to proposition (1).

If schools are essential for disciplinary education, we must allow them to do the work for which they are essential. Two necessary factors are able teachers and willing students. A shortage of the former makes it imperative not to overload them with unwilling students.

What of the unwilling students; should we let them go and forget about them? Not necessarily; but perhaps we should accord some right of freedom of choice even to an adolescent. Some want academic learning. Others want other kinds of learning. Others just want to live, and perhaps to learn a little, as they go along. Have we lost all faith in freedom? I think not. We already have many alternative pathways to education which provide a fairly rich choice. We can make it richer. Disciplinary education other than the conventionally academic is provided by a variety of institutions, namely vocational schools, art schools, professional schools and various schemes to learn while you earn. Should we not expand them and increase them? They can and should provide healthy competition for conventional academic institutions. These are the facilities which need to be developed to maximize opportunities for disciplinary education with freedom of choice.

For those who elect none of the disciplines, I throw forth the suggestion that there are still the mass media of communication. How powerful they are needs no emphasis. Who can escape them? Surely here is the challenge of the present and future, and to no one more than to the newly formed Canada Council.

In summary, it may be stated that in Canada as elsewhere in the Western world, people are living longer, presumably as a result of improved control of physical diseases. Mental ailments and moral delinquency, however, continue to increase and constitute the biggest problem facing the medical and social sciences. Increased life-expectancy and material prosperity have been accompanied by unprecedented demands for education for everyone. Universal primary education has been and can be supplied in standard packages. Demands for universal secondary and even higher education threaten to swamp the facilities and debase the product. Diversification of channels for post-primary education should be a practical aim, and maximum freedom of individual choice should be an ideal. Mass media have contributed much in our time at the evocative level of education. Perhaps they can do even more.

L.-P. DUGAL

THE BRITISH HISTORIAN, H. G. Wells, has written: "Human history becomes more and more a race between Education and Catastrophe." If then, and I think that we all agree, the very survival of any democratic country, including Canada, is conditioned by education, I do not know of any better way of paying "our debt to the future" than to ask ourselves now, as others have done for us before, what are exactly the aims and purposes of education; what should we do to fulfil them as completely as possible so that our children and grandchildren may be at least a little better than we are, and the world in which they live, a little brighter.

In the few minutes at my disposal, I have no time to wonder why such a far-reaching task has been committed to one of those so-called specialized university teachers, unless it is to make at least one of them realize that the process of focusing our mind on a very limited field, the fact of being deeply immersed in our observations and experiments, although a necessary process, may cause a disproportionate view of our world; such a situation is liable, if we are not careful, to make even teachers lose sight of the true aims of education; that is—if we ever had a clear understanding of them. We realize with great surprise and dismay that we could be classified as fanatics, since a fanatic has been defined as "a man who redoubles his efforts when he has forgotten his goal."[1] It does not seem then to be a waste of time to examine closely what are those aims and purposes of education and what efforts we can make, all of us now, professors and teachers included, to apply them to the greatest benefit of the future generations.

As you probably all know, or at least have heard or read some time during your college days, or later in life, "Education should aim at training the mind and at developing it beyond the inculcation of certain basic skills; it should not be vocational training."[2] It must teach people how to think, not what to think. In other words, it should not be dogmatic; it must be a tool that leads the student to question our culture and to desire and effectively to acquire culture, and not produce satisfied bachelors that think that they are educated because they know what to think; it should enrich the human spirit, impart a love of knowledge and wisdom, develop intellectual independence, rather than produce an accumulation of factual knowledge; it should help people to define themselves and their fellow men. In summary, education must aim at developing, in an appropriate way, the human characteristics of this complex unit that we call a human individual.

Some practical people will object that these aims, these principles, are too ideal, too theoretical, and that they do not take into account the immediate needs of our country and the needs that are the most urgent for the near future. Of course, the community should have its doctors, its lawyers, its scientists, its engineers, but there is nothing to prevent those people from being trained according to the best principles of education. Even if the need for manpower in the fields of science and engineering is the most urgent at the present time, let us not go into mass production at the expense of quality. Incidentally, let us not forget that industry is not looking merely for diplomas, but for humans with a solid grasp of fundamentals, with analytical power, and with capacity to adapt themselves—adaptability being a characteristic

---

[1] Quoted by Hans Thirring, *American Scientist*, XLIV (1956), p. 445.
[2] G. Kirk, *A.I.B.S. Bulletin*, V (1955), p. 1.

of living matter—constantly to new problems and to solve them. One must not forget that the responsibilities of life never look exactly like the problems presented at school, but that the gymnastics of training the mind to solve problems at school will enable that mind to be in a fit condition for solving the more complex and ever changing problems of the terrible "daily."

I said a moment ago, referring to the need for scientists and engineers, that we should not sacrifice quality to quantity; we must put "our emphasis on goods of fine substance at high price rather than of moderate quality at low cost."[3] If we aim at quantity, the quality may deteriorate so much that we get nothing; if, on the contrary, we aim at quality, we have a slight chance of having also the quantity, but at least we will produce leaders and that is what we need, now, tomorrow, always. Our greatest responsibility, vis-à-vis the generations to come, is to ensure that our system of education will produce leaders, flag-officers rather than people discussing flags; leaders who are intelligent people, men of knowledge, imagination, high moral character, integrity, sincerity and wisdom.

Now, one might wonder how an educational system which, step by step, from the primary school to the university, is intended to produce leaders can include and take care of the rest of the population? I think that the mass of the population would simply possess the same qualities as its leaders, at degrees varying with the level of intelligence and intellectual power, and would thus be equipped to follow genuine leaders that they would be prepared to admire, and to turn down false prophets. Let us never forget that no man can live without admiration: let us take care that those who excite that admiration owe it to their education and to the education of their admirers, and not to the lack of it on either side or on both.

The question remains now as to what educational system or systems should be entrusted to attain those aims that I have outlined. Here we come to a very controversial, yet necessary question: is it true that the liberal arts schools are the sole custodians of the liberal spirit? Or is it true that schools of science have the only sound approach to education? I suppose you all know that in February of next year, there will be here, in Ottawa, a Conference on Education which will be attended by the most distinguished educators of this country; and I am sure they will discuss this matter of the relative values of humanities and science. In the meantime, it will not hurt or impress anybody

[3]J. A. Stratton, *A.I.B.S. Bulletin*, VI (1956), p. 1.

if I myself venture to say that I am in favour of liberal arts education, provided that it does not reject or disdain science and technology. As a human being, I believe in the value of science as a means of mental discipline. Biology, for instance, and I quote Julian Huxley, "has a large variety of the universe as its subject matter; its position is midway between the physico-chemical and the social sciences. Its value as a branch of discipline is partly that which it has in common with all sciences—the training and strengthening of common sense; partly that which is more peculiar to itself—the great exercise which it affords to the faculties of observation and comparison; and the exactness of knowledge which it requires on the part of those among its votaries who desire to extend its boundaries."

I think this is a good time and a good place to make a plea for the teaching of some biology to all students at the high school level; when one is confronted, as we were earlier in the symposium, with the penalties of ignorance of man's biological dependence, and when one stops a moment and thinks about the fact that man is body and soul, it is surprising and even scandalous to think that so many high school graduates never have a taste of the sciences of life.

If I believe in science because I am human, I also believe in humanities because I am a scientist. I believe that scientists and engineers should have the humanistic quality attributed to the Greeks, that they should be refined, superior, complete human beings and not only super technicians. Hence, restoring the proper balance between liberal arts and science, namely uniting modern knowledge, seems to be one of the main responsibilities of our time towards the future. Such a unification would imply, of course, a pattern of education sufficiently flexible, especially in the early formative years, to satisfy later both the needs of the nation and the aspirations of the individuals.

And so, if we keep in mind the aims and purposes of education, if we convince ourselves that for their fulfilment we must be, as Floyd Zulli says of André Obey in his *Noé*, "old-fashioned enough to believe in spiritual values, in beauty, poetry, and above all, in humanity" and if, I would add, we always keep in mind that "bonum et verum sunt unum et idem," that which is good and that which is true are one and the same, then we may hope that with God's help, man, far as he is from being perfect, may not only survive but be, some day, just a little bit better than he is now.[4]

[4]Other sources are: J. R. Killian, *American Scientist*, XLIV (1957), p. 115; and R. W. Gerard, *A.I.B.S. Bulletin*, VI (1956), p. 20.

W. B. LEWIS

LIKE MOST PROBLEMS worthy of discussion, the efficient use of manpower has two aspects; one in this case is happy, the other unpleasant. Dr. Steacie in his paper has posed the question, "how much efficiency do we want?" That causes us to think of the unpleasant aspect. You may consider that idyllic enjoyment is more attractive than efficiency; but why make the distinction? Enjoyment is indeed the happy aspect of efficiency. I take it that we all recognize that work is a necessary part of our debt to the future, and we are moreover likely to claim some purpose for our own work. Few, however, would claim to achieve their purpose without suffering frustration. The happy aspect of the efficient use of manpower is just this minimizing of frustration in accomplishing our purpose, or, in shorter words, getting rid of the blocks in gaining our ends.

Sometimes it helps to look squarely at the blocks. A slave helping to build the pyramids in Egypt may have thought his only choice was between the whips and the burden. A university professor of today may feel he has to keep peace with the dean and keep pace with the scripts of the students. If the slave could have acquired a bulldozer, or the professor a—well, perhaps in the year 4000 A.D. there may be a name for it. If I may quote some well-known lines from Lowell,

> New occasions teach new duties
> Time makes ancient good uncouth;
> They must upward still and onward
> Who would keep abreast of truth.

The slave may have said, my father always told me to keep in the middle of the gang, for it is those on the outside who get hurt. The professor may rely on maxims from his professor. Let us take some samples: "chemistry students must learn to pipette," "a physics student must be well grounded in conic sections," "a radio-engineer must first be an electrical engineer"; or some more advanced maxims: "the purest mathematics are those of no possible use," "laboratory work should be done between 9 A.M. and 6 P.M.," "students should not own more books than they can read," "any man worth his salt should be able to do research on $2,000 a year"; or still more subtle: "he is only a second-class man, he can expound but not create," "he is only fit for school-teaching," "he is a pure mountebank good enough for the elementary classes"; and lastly, "the student must have intimate contact with professors engaged in research." Many of these are no better than the slave's father's maxim; good formulae for the treadmill but hopeless for radical progress.

The real problem facing the universities today is that many more students than formerly hope to pass examinations but professors do not exude wisdom notably more rapidly than their predecessors. Indeed, the professor who now maintains a position in the forefront of research is likely to find it demands even more of his time. Thirty years ago it might have been possible for a qualified physicist to take up a new field and do competent research within six months. Now it takes that time to read the literature to discover the frontier, which is moving so rapidly that with part-time reading it may never be reached. New ways must be found quickly.

There is a great temptation for a specialist, who by a brilliant intellect and hard labour has made himself master of an intricate subject, to regard himself as a high priest through whom only a limited number of elect students can be initiated into the mysteries. This is dangerous. Another master may, rather, regard himself as a slave of the temple, keeping it polished for all to derive the glimpses of truth they can absorb. The latter is more efficient. The problem is not an easy one; it has several aspects.

I am speaking against the high priest, but we cannot dispense with what made him a high priest. As Milton wrote,

> Fame is the spur that the clear spirit doth raise
> (That last infirmity of noble mind)
> To scorn delights and live laborious days;

Milton was concerned with the moral ending,

> Comes the blind Fury with th' abhorred shears
> And slits the thin-spun life.

But in the present context we must recognize and approve this spur to live laborious days; although it is morally regarded as a human weakness, it nevertheless yields fruit. The high priest must not be deprived of his honour, but the mass of students must be able and encouraged to honour the temple slave so that in his turn he may experience the spur to live laborious days.

Much can be done by exploiting our means of communication. In the old days the research professor, the high priest, published his work in scholarly fashion. At rarer intervals, the essence of the published work would be brought together by him or another in a standard text. Nowadays to get the essence distilled and disseminated, we have travel, conferences, reviews, journals, films, radio and television; but how poorly they are exploited in serious scholarship and how inadequate is our performance. When we laughed at that quip, "The vision we have is television," we were admitting our failure so far to raise television to a higher level. Could we not find some fame for the temple slaves, the pioneer producers of accessible documentary films of real scholastic worth? I feel sure the ways and means will be found by ingenious pioneers. We are their greatest obstacle with our narrow traditional views.

We have lived through some revolutions of thought; we can look for more to come. Personally I had the good fortune to start research at the Cavendish Laboratory in Cambridge just two years before the marvellous year of 1932 that saw artificial nuclear disintegration and the neutron discovered. For our research Rutherford acquired a specially designed magnet which cost £250, an amount that then required some apology. Rutherford said, "Yes, it is a lot of money, as much as a research student for a year, but it will do a lot more work." Five years later we were building a cyclotron for £20,000 and Rutherford was receiving £500,000 for a great expansion of the laboratory. Now at Chalk River we have reactors valued at $15 and $55 million that are not even to be regarded as luxuries; that is to say that by their work all this expenditure promises to bring within twenty years a direct manifold return. This marks the great disparity between the scale of equipment in universities and that in industrial or government-supported research institutes. The cost per man in the latter is assessed and adjusted in the interest of maximum productivity. This cannot, of course, be done by exact arithmetic; one of the most intangible aspects is the atmosphere that breeds creative thought. For this I happen to believe that mental pressure and stimulus apply a better spur than physical hardship and I seem to find support for this from industry and departments of education. In universities there are two other stimuli, namely the need for clear thinking in order to teach and the constant impact with fresh young minds. I do not believe that either the university or the research institute should be destructively critical of the other, but they can and should be mutually helpful.

In the research institute we estimate that the cost per professional scientist lies between $20,000 and $30,000 a year, that is to say over three times but less than five times his salary. The indications are that for greater productivity this ratio should rise, and not by paying the scientist less. This is part of the trend towards automation in industry. As Louis Ridenour has remarked, "There is in a very real sense no competition between man and automatic control systems. There is no competition because men are outclassed from the start." We find ourselves wrestling with major problems of data processing as we make our apparatus automatic and process the results with electronic digital computers. The role of the man becomes to judge, organize and control the material which the machines handle. In an advancing field this material is continually changing.

To take some very simple examples—we would have to pay more for messengers to take books to and from a central library than to provide copies to the scattered laboratories. Library books need good bindings for preservation, books containing data mostly of transient value have to meet a different specification for laboratories. Conditions in universities are somewhat different again.

We may note also that our major equipment such as the reactors and our Van de Graaff accelerator operate either for 24 hours a day or for at least two shifts. Most scientists also work much more than the standard week honoured by labour. This, I believe, is efficient, not perhaps per man-hour, but by the standard of accomplishing our purpose.

That is the happy aspect of efficiency, opposed to all restrictive slogans and the "holier than thou" thoughts of high priests. Now we must turn to the less pleasant aspects where curbs and restrictions are a necessary part of our debt to the future. To present this briefly I will take advantage of thoughts already left with you from previous speakers. For example, the fishermen of the Maritimes. Their life is hard and hazardous. If any new industrial activity were to be introduced with anything approaching their mortality figures the moral outcry would be overwhelming. But fishing is an occupation of long standing to which men are attached regionally and romantically. If we improve the efficiency of the fisherman we run up against the limits of the market and of the resources. The economist would then show the wisdom of reducing the number of fishermen. This is something that can be done by making available to young men and women of the region more attractive jobs and not boosting the economically inefficient industry from taxes levied on the more productive industries. This aspect is undeniably sentimentally unpleasant, but to anyone believing in the sanctity of human life, as well as to the hardened economist, it is surely part of our responsibility to the future to stimulate the change.

In this symposium it has been lamented that the native Canadian is playing a relatively small part in the hazardous adventures of developing the Canadian north. For those wise enough to know the value of the life of hazard the opportunities exist.

As scholars and scientists, our debt to the future is to avoid fomenting the glee shown by this audience, as well as by almost all readers of fiction, especially of detective novels, in pitting man against man. That is laudable here in debate but highly dangerous in action over the whole range from domestic relations to atomic war. For efficiency, there must be sparks from friction in debate—the conservationist must meet the economist, the humanist the scientist—but there is danger in extending this friction to the life of action and it wastes manpower.

I spend a great deal of my time preoccupied with another problem. Economists recognize that a close correlation exists between productivity, or Gross National Product, and physical energy consumption. In Canada we are just entering a phase where we need to determine where our future energy supplies are to come from. Economic hydro-power will soon be fully harnessed. The demand is rising with our Gross National Product. Physicists have discovered how to release a vast store of energy from our resources of uranium. We now believe this can indeed be obtained competitively with power from other sources in many industrial areas. We have, however, to use our manpower efficiently because large expenditures are required to establish the engineering aspects.

To return to the university problem, we have seen the predicament of the professor, where, by the established ways of life, both research and students are making more pressing claims on his fixed allocation of time, hours per day and working life. We have recognized some of the contributions towards satisfaction in his life. First, the human spirit spurred by forces of high or low moral repute. How he selects and recognizes his purpose is a matter for the individual and his environment. When this purpose is recognized the spur may lead to overwork or overworry. Here the modern world can offer aids from the typewriter to the digital computer, the microphone to television and the fantastic range of copying of complex patterns offered by the photographic art, applied from the electron microscope to the cinema. This mode of life demands a framework of a stable economy and an adequate supply of material energy. The power to achieve and the limitations are set by man.

I would like to end by quoting a well-known prayer, "Lord grant us grace to accept those oppressions which cannot yet be changed, strength to struggle with those that can be changed, and wisdom to know the difference."

Lightning Source UK Ltd.
Milton Keynes UK
UKHW020022210722
406167UK00009B/766